PRAISE FC

SOLD DOWN TH

'The forensic detail that the authors g
A really, really grea… book.'
VIRGINIA TRIOLI, ABC RADIO MELBOURNE

'[A] deeply disturbing book...For the river system—and the people dependent on it—the market brought nothing but catastrophe.'
JEFF SPARROW, SATURDAY PAPER

'If you're a fan of US author Michael Lewis...
you will devour this magnificent book.'
PETER DONOUGHUE, BOOKNOTES

'This book is essential reading.'
BOOKS+PUBLISHING

'A vitally important book detailing how a policy meant to save Australia's rivers became the means of their destruction.'
TIM FLANNERY

'A complex subject made enjoyably readable. *Sold Down the River* shows how water has become commodified in the tentacles of mercenary traders—and at a time when climate change is making water more important and more scarce.'
CHERYL KERNOT

'A compelling exposé of how market forces are destroying the Murray–Darling Basin. The tragedy is we let it happen.'
MICHAEL CATHCART

'It takes writers with excellent knowledge, uncommon flair and impressive skills to get to the dark truth of Australia's epic water failure. Read this book and weep—better still, read it and demand change.'
ROSEMARY SORENSEN

Scott Hamilton is an expert in natural resource management and climate change. A member of the Energy Transition Hub at the University of Melbourne, he writes regularly for the *Mandarin* and other publications.

Author and historian Stuart Kells is adjunct professor at La Trobe Business School. He has twice won the prestigious Ashurst Business Literature Prize.

SCOTT HAMILTON
and STUART KELLS

SOLD DOWN THE RIVER

How Robber Barons and Wall Street Traders Cornered Australia's Water Market

TEXT PUBLISHING MELBOURNE AUSTRALIA

The Text Publishing Company acknowledges the Traditional Owners of the country on which we work, the Wurundjeri people of the Kulin Nation, and pays respect to their Elders past and present.

textpublishing.com.au

The Text Publishing Company
Wurundjeri Country, Level 6, Royal Bank Chambers, 287 Collins Street,
Melbourne, Victoria 3000 Australia

Copyright © Scott Hamilton and Stuart Kells, 2021

The moral right of Scott Hamilton and Stuart Kells to be identified as the authors of this work has been asserted.

All rights reserved. Without limiting the rights under copyright above, no part of this publication shall be reproduced, stored in or introduced into a retrieval system, or transmitted in any form or by any means (electronic, mechanical, photocopying, recording or otherwise), without the prior permission of both the copyright owner and the publisher of this book.

First published by The Text Publishing Company, 2021
Reprinted 2021

Cover design by Jessica Horrocks
Cover images by Jes2ufoto /Alamy Stock Photo and iStock
Page design by Rachel Aitken
Typeset by J&M Typesetting
Map by Simon Barnard

Printed and bound in Australia by Griffin Press, part of Ovato, an Accredited ISO AS/NZS 14001:2004 Environmental Management System printer

ISBN: 9781922458124 (paperback)
ISBN: 9781922459459 (ebook)

A catalogue record for this book is available from the National Library of Australia.

This book is printed on paper certified against the Forest Stewardship Council® Standards. Griffin Press holds FSC chain-of-custody certification SGSHK-COC-005088. FSC promotes environmentally responsible, socially beneficial and economically viable management of the world's forests.

If the misery of the poor be caused not by the laws of nature,
but by our institutions, great is our sin.

CHARLES DARWIN

Whisky is for drinking and water is for fighting over.

ROBERT BUCHAN, MAYOR OF ST GEORGE, QUEENSLAND,
IN *TWO MEN AND A TINNIE*

CONTENTS

PART IV: CORNERING THE MARKET

PART V: FIXING THE MARKET

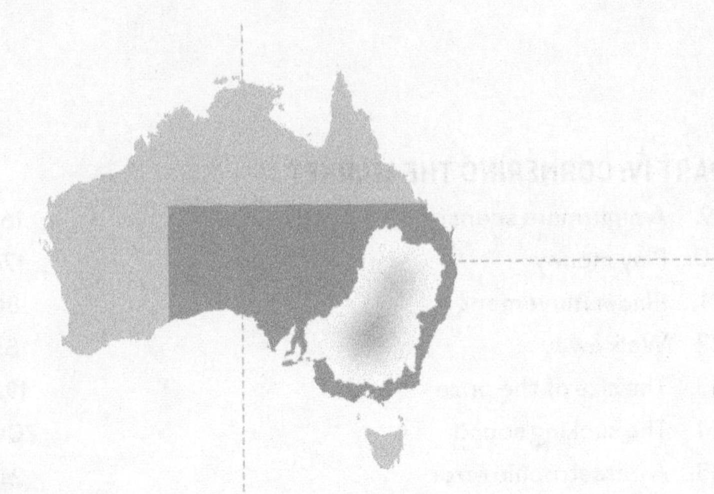

SA

ADELAIDE

The MURRAY–DARLING BASIN

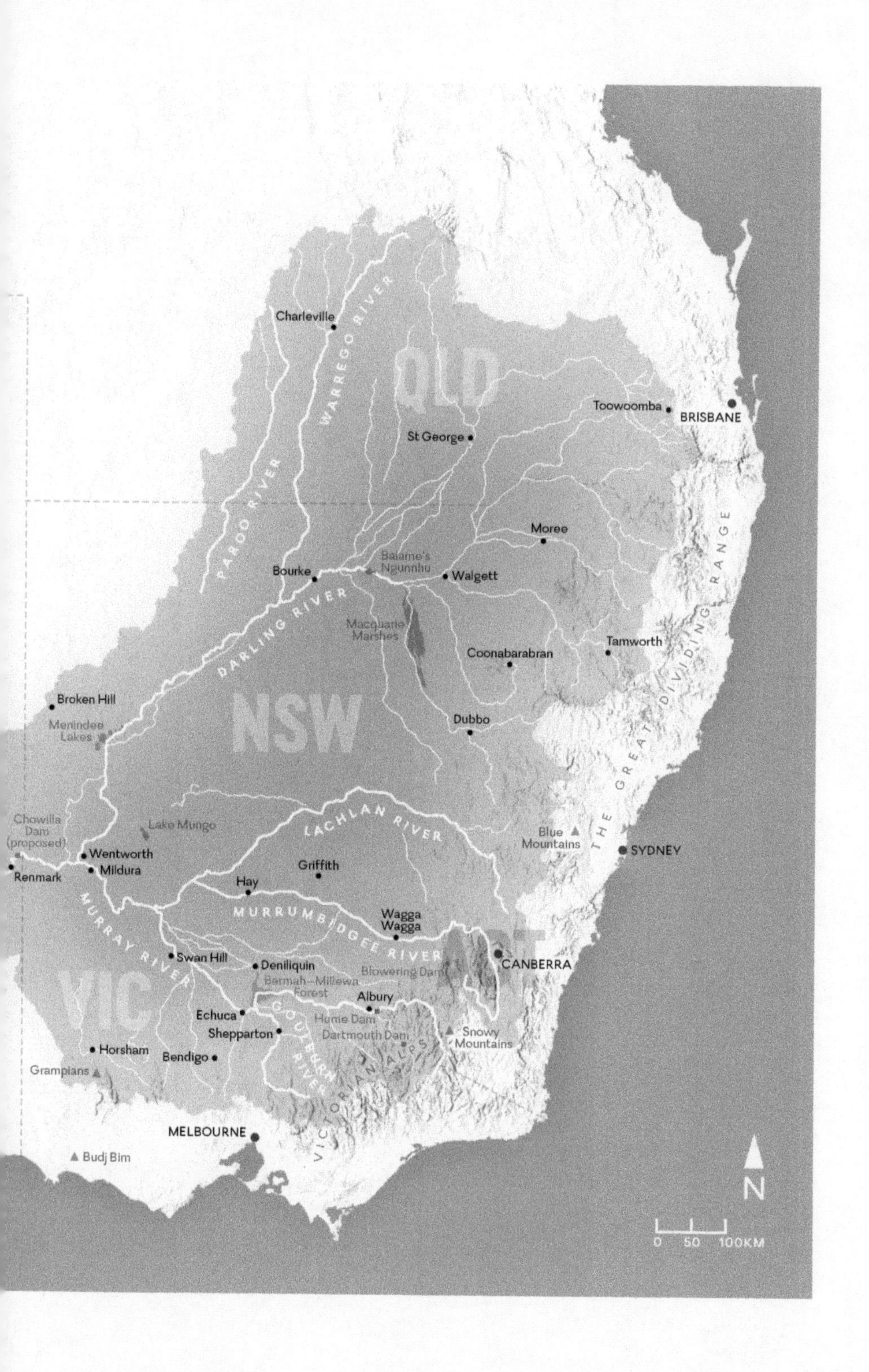

Charleville
WARREGO RIVER
QLD
Toowoomba
BRISBANE
St George
PAROO RIVER
Moree
Bourke
Baiame's Ngunnhu
Walgett
DARLING RIVER
Macquarie Marshes
Coonabarabran
Tamworth
THE GREAT DIVIDING RANGE
Broken Hill
Menindee Lakes
NSW
Dubbo
Chowilla Dam (proposed)
Lake Mungo
LACHLAN RIVER
Blue Mountains
SYDNEY
Wentworth
Mildura
Renmark
Griffith
Hay
MURRUMBIDGEE RIVER
Wagga Wagga
MURRAY RIVER
CANBERRA
Swan Hill
Deniliquin
Blowering Dam
Barmah–Millewa Forest
VIC
Albury
Echuca
GOULBURN RIVER
Hume Dam
Shepparton
Dartmouth Dam
Snowy Mountains
VICTORIAN ALPS
Horsham
Bendigo
Grampians
MELBOURNE
Budj Bim
N
0 50 100KM

PROLOGUE

Lachlan Marshall, Riverina farmer: There are lots of stories you'll hear and people you'll talk to as you delve into this. You are about to go down a rabbit warren that will scare the living daylights out of you. The best way I've had it explained to me is by a person I know in finance, who lives in an irrigation area. That person said the water world is so full of profiteering and white-collar crime that the mafia in Griffith wish they'd got into it in the beginning, but they missed the boat.

The more stories you hear, the more you'll see how inside knowledge is used—and how rules are flexed and broken to achieve set outcomes. Honestly, you're going to embark on a journey where you will daily say, 'Oh my God, I cannot believe this!' And you'll think back to me telling you this—because of the effect that it's had on the community.

I remember all the people who've gone. This region was created as an irrigation region. Before irrigation it was all massive sheep stations. And then, post-war, it was a nation-building thing to bring everyone here, through the Depression. Then we had settlement and it became the foodbowl to the nation.

That is being undone. Water has gone from hundreds of dollars per megalitre to thousands of dollars. The amount of irrigation that happens is probably less than half of what it was. Someone is making a killing, but not the farmers. Water trading has destroyed our community. It's gutted it. Shelled it.

INTRODUCTION
WHY LOOK AT THE WATER MARKET?

How did we come to write this book? What led us to take a close look at how Australia manages its scarce water resource?

A few years ago, we began researching political bipartisanship in Australia, as part of a project at Melbourne's La Trobe University. We interviewed politicians to find examples of good and bad decision-making, with a focus on how different political parties could come together and make good policy.

One example of bipartisanship—and, more recently, a counter-example—kept coming up in our interviews: the Murray–Darling Basin Plan.

The Murray–Darling Basin is a water catchment that consists of 77,000 kilometres of rivers and streams. It covers more than a million square kilometres—an area larger than France. Home to more than 40 Indigenous nations, it encompasses most of inland

New South Wales, plus most of inland Victoria, a sizeable part of Queensland, most of the arable areas of South Australia, and all of the Australian Capital Territory.

In 2012, the four states and the ACT signed up to the Murray–Darling Basin Plan as a way to manage scarce water across the Basin. The Plan specified how much water could be extracted from rivers, lakes and creeks for irrigation. It also allocated water for the environment, so natural ecosystems could be protected.

We looked into the Plan, and our first major article on Basin bipartisanship appeared in February 2019 in the national public policy magazine, the *Mandarin*. 'Water,' we said, 'provides many examples of how politics can get in the way of good policy, and how Australia's federal system sometimes supports and sometimes derails bipartisan solutions.'

The success or otherwise of the Plan is crucially important. An area of great social and cultural significance, especially for Aboriginal people, the Basin is home to almost 3 million residents and accounts for two-fifths of Australia's total food production, including fruit, milk, beef, pork, chicken, rice, wheat, olives and almonds. The total annual value of Basin agricultural production is more than $24 billion.

The Basin is also of worldwide ecological importance, and includes 16 Ramsar-listed wetlands. Australia is a signatory to the 1971 Ramsar Convention, which aims to halt the loss of wetlands around the globe, and to conserve those that remain. These 16 wetlands are deemed to have international ecological, botanical, zoological, limnological or hydrological significance. They place the Basin at the sharp end of concerns about biodiversity and climate change.

—

We continued to study the Plan and its central feature, the 'cap and trade' approach to managing Basin water. We wrote further water-related pieces for the *Mandarin* and the University of Melbourne's *Pursuit* magazine. These articles opened a lot of doors and mouths, especially when we began to think something might be wrong at the core of our water market, which happened to be Australia's biggest experiment in natural resource management policy.

The water market is large and growing. The trade in Basin water rights (permanent plus temporary) is now around $1.8 billion per year. The total value of Murray–Darling Basin water rights is estimated to be $26.3 billion. Water accounts for around 40 per cent of the value of Basin farms' assets.

Farmers, brokers, bankers, regulators, market operators and even some professional water traders were eager to talk to us and to tell us what was going on in the market. The insiders we spoke to felt that few outsiders were interested in understanding the reality of the market, and that most outsiders they'd previously spoken to—especially those from academia and government—weren't really listening at all, or were asking the wrong questions.

So we made a lot of calls and followed a lot of leads. We asked different questions. We listened. And, unlike many others who had come before, we helped put the puzzle pieces together.

To understand what might happen at the junction between modern markets and ancient water, we spoke to people with direct experience of the intersection between finance and computing (aka 'FinTech'). Most important of all, we spoke to people from Basin communities, including Indigenous and non-Indigenous people in towns and on the land. The farmers we spoke to represented a wide variety of different types and scales of farms.

To some degree, we had a head start. The water market is a

multi-disciplinary phenomenon, and we brought a multi-disciplinary perspective. Between us, we'd worked in banking and finance, academia, energy, engineering, public policy, environmental science, education, regulation, consulting, auditing, politics, publishing, and journalism—under our own names as well as ghost-writing for some of Australia's leading financial journos.

For our own writing, we'd researched a range of subjects, including market design, climate change, short sellers, privatisation, anti-corruption agencies and the Hayne banking royal commission. Our collective work in finance extended to advising the Australian Bankers' Association on the 1997 Wallis banking inquiry, and working with the liquidators of Lehman Brothers Australia after the global financial crisis.

Another potential advantage: we had both lived in regional Victoria and New South Wales—in places such as Albury, Bendigo, Wagga Wagga and the Latrobe Valley—and we still had strong connections to those regions and nearby ones. So to some extent we already understood rural issues and could speak the same language as people on the land and in regional towns.

But more important than any rural charm or regional credentials we might have had, the farmers and other people in the water market were determined to get their stories out, and they hoped they'd found two people who might listen and understand how water trading actually worked, and what it felt like on the ground.

Through chains of contacts and farmer introductions, we tapped into an informal communication network, part bush telegraph, part dark web: the Australian water market underground. At pubs and clubs, on farms and at field days, and via private emails and apps, the undergrounders swapped market trends, trading strategies, rumours, scuttlebutt and conspiracy theories about who was winning and losing, and who was doing what to whom in the market.

The dark water conversation was heavy with jargon and codewords, and much of it was impossible to follow. The jargon is part of a bigger problem with water words. Water regulators and officials speak their own language (actually several languages), and words are often used to confuse insiders and exclude outsiders. As Lynne Kirby observed in 2017 about the Californian water system, 'It's byzantine…a priesthood. It's deliberately obfuscatory language. They are trying to keep people from figuring it all out. It is a shell game…Things have multiple names and they all sound the same. The "Department of [this]". "That water board…" Blah, blah, blah.'

In Australia, the different Basin states define water terms differently—a new version of the old rail gauge problem. Adding to the confusion, people who've come to water from different professions and academic disciplines use the same words to mean different things. Hotly contested words such as 'rights', 'efficiency', 'investment', 'speculation', 'allocation', 'access' and 'liquidity' are some of the worst sources of misunderstanding.

We built a lexicon to untangle and decode these terms. This customised resource became our compass as we went further down the rabbit hole of Australian water trading. With the help of this tool—and our trusted guides in the dark water network—we began to understand.

We came to know the nature and history of the river system—how Basin water had been managed and used, going back to a time well before the arrival of Europeans. We understood how the water market worked at a practical level, and how Australia had arrived at the current point, where many farmers lacked confidence in the market, despite the assurances of a generation of politicians and policymakers.

And at the heart of it all, under all the jargon and theories and estimates and plans, we uncovered a catastrophic error, a broken promise, a market failure.

This book is the story of what we found.

PART I: ORIGINS

CHAPTER 1

THE NOAH EFFECT

Australia is the driest inhabited continent in the world. About half of our inland-flowing rivers never reach the sea, instead evaporating in ephemeral salt lakes, or just ending nowhere in particular. But the largest rivers that flow west from the Great Dividing Range eventually reach the Great Australian Bight.

Principal among those rivers is the Murray, originally called Tangula by the Yorta Yorta people. That river bends and skirts over a distance of 2,508 kilometres until it reaches the lower lakes in South Australia, and ultimately the sea. Five of Australia's next six longest rivers—the Murrumbidgee, Darling, Lachlan, Warrego and Paroo rivers—also rise in the Great Dividing Range and all eventually merge with the Murray. Together with their tributaries, these watercourses form the river system of the Murray–Darling Basin—the only major river system on the continent.

Mountains are bridges between geology and climate. The Great Dividing Range, formed around 130 million years ago, creates microclimates of clouds, rain and snow, which give rise to rivers and creeks. It is by far Australia's largest and longest range: it stretches 3,500 kilometres, from Dauan Island in the Torres Strait to Mount William in western Victoria's Grampians. Along the way, it forms the Blue Mountains in New South Wales, and the Australian Alps, which straddle New South Wales and Victoria.

The Alps consist of two subregions: the Victorian Alps and the Snowy Mountains. The Snowies include Australia's highest mainland peak, Mount Kosciuszko, which reaches 2,228 metres above sea level at its summit. This is far below the heights of the Himalayas (8,849 metres) and the European Alps (4,810 metres), which are tall enough not only to make their own weather but to maintain numerous year-round glaciers. In Australia, however, glaciers were formed only in the vicinity of Mount Kosciuszko, and they did not survive the end of the last ice age.

The Australian Alps, in short, are not very alpine. In summer, temperatures often reach 30 degrees on the highest peaks. The annual mountain runoff—around 23,600 gigalitres into the Basin rivers—is meagre by world standards. By way of comparison, the runoff into the Mississippi River catchment is more than 20 times larger, and the Amazon River's flow is much larger still. More water flows in the Amazon in a day than in the Murray in a year.

In 1856, about halfway between the Australian Alps and the South Australian border, Joseph Waldo Rice founded the town of Barmah on the Victorian side of the Murray River. The town's name comes from a Yorta Yorta word that means 'place of the plover'.

The town is next to the Barmah–Millewa Forest, which boasts

the world's largest concentration of river redgums. In the 19th century, men from Barmah cut the timber into sleepers, then hauled them on bullock trains to Echuca, where they were loaded on paddle-steamers. This enterprise—along with fishing—was the economic foundation of the town.

Today, if you drive south along the Cobb Highway between Deniliquin and Echuca, you will see to the east a striking feature in the landscape: a long, earthen embankment that almost looks man-made. But it isn't: the feature is the sandy lip of the Cadell Faultline. Some 25,000 years ago—recent enough to be described in Dreaming stories—a series of tectonic uplifts along the faultline raised a triangle of country between Deniliquin, Echuca and Swan Hill.

As a result of that upheaval, a 72-kilometre stretch of the Murray River's bed rose as much as 40 feet (12.2 metres). The river's flow was severely disrupted. Some of the water backed up to form a natural lake. Some water went north, around the obstacle, to create what is now called the Edward River anabranch. But most of the water went south, forming the Murray's current channel and the Barmah Choke.

Literally a choke-point in the Murray, the Choke is a narrow and shallow section of the Murray that runs through the Barmah–Millewa Forest. It determines and constrains the flow of the entire river, and the flow of water in this section is limited to about 7,000 megalitres, or megs, per day. (In recent times, the capacity has fallen to that level from around 10,000 megs per day, mainly due to the accumulation of silt.) If the flow in the Choke exceeds that amount, water spills out of the river and into the adjacent lakes and wetlands, and into the forest. Such spillages happen often and are clearly visible as brownish blooms on Google Earth.

—

The Basin flows are highly seasonal and often unpredictable. Over the years, to predict rain and flows, farmers relied on science and the weather bureau—and they relied on local lore. Farmer Shelley Scoullar shared anecdotes from various farmers, such as this one from a dairy farmer in Cohuna: 'Ants building their nest walls high is a sign of a wet season.' Another farmer agreed: 'We have found that the ants are never wrong when a storm is coming.'

Farmers also looked to birdlife. A mixed farmer from Finley: 'My dad always said when the laughing chooks (aka kookaburras) laughed, rain was coming!' Another farmer: 'Ducks nesting in the old trees at home. Big rain coming.' A Deniliquin farmer said owls, too, were a sign of rain, and we also heard that currawongs coming out of the hills to lower ground signalled the approach of a big storm.

Quadrupeds could also predict rain. 'My mum is convinced,' a farmer said, 'that if she sees a horse lying down in a paddock, it will rain two days later.' Cattle, too, were known to lie down before a decent rain. 'We've always thought,' another farmer said, 'that this was because they know once the rain starts they most likely will be standing for some time until it stops again.'

Climate change has made the flows even harder to predict. In 1980, Australian *Playboy* magazine published an in-depth article titled 'The Greenhouse Syndrome', in which Richard Shears quoted Australian scientists and farmers who spoke to him about climate change and water scarcity:

> In a good year on Tallagandra Farm, dairyman Terry Makin's 200 Friesians earn him $84,000 in sales of milk and butterfat. If the rain doesn't come, Terry loses

> $14,000 in sales and has to spend an additional $7,000 on hay and fertiliser.
>
> 'What I want,' he says, 'is a bloody good weather forecaster.'
>
> He needs a farmhand like Joseph, that Biblical prophet who interpreted the Pharaoh's dream and warned that seven good years on the land would be followed by seven years of famine. Unhappily for farmer Makin, after 3700 years of industrial progress, the greatest scientific brains in Australia and the rest of the world are still guessing at next week's weather.

An important concept in finance and investing is based on the same Biblical episode that *Playboy* mentioned. Fractal mathematician Benoit Mandelbrot coined the terms 'Joseph Effect' and 'Noah Effect' to describe how movements in asset prices and markets tended to be parts of larger trends and cycles. (Specifically, the 'seven good years' are known as the Joseph Effect, while the 'seven years of famine' are the Noah Effect. Scarce Australian water is much more Noah Effect than Joseph Effect.)

One example of a larger pattern is the approximately seven-year business cycle that is sometimes observed between economic booms and busts. (Another is the regular cycle of major water policy reviews.) Mandelbrot's idea was a powerful challenge to the orthodoxy that asset price movements followed a 'random walk', such that the best way to predict future values was to look at current ones.

Applying the mathematics of fractals and chaos involves searching for hidden complexity and recurring patterns at different scales and over different time periods. Mandelbrot and other fractal thinkers such as Edgar Peters would search for fractal harmonies in commodities and securities markets. Economic modellers and climate modellers would adopt similar techniques when searching for fractal and chaotic patterns to explain the mysteries of the economy and the climate.

CHAPTER 2
ANCIENT, STONE-BUILT WATER INFRASTRUCTURE

For 70,000 years or more, people have lived on the Australian continent. Aboriginal and Torres Strait Islander peoples adapted to its ecology and climate, building a rich tradition of art and culture and assembling a deep body of knowledge about the land and natural resource management. This body of knowledge is science and engineering as well as culture and tradition.

Water was integral to the survival of Indigenous peoples, for whom rivers and creeks have long had great significance. Indigenous maps, language boundaries, Dreaming stories and ceremonial places are frequently based on or oriented towards water. According to the Australian Institute of Aboriginal and Torres Strait Islander Studies, Indigenous peoples' relationship with the land 'involves an ancient history with the land and unique systems of law, custom and spirituality that regulate the land and water management'.

In the Murray–Darling Basin, there has been growing acknowledgment in recent decades of the richness of Indigenous heritage, the rights of Indigenous people to water, and the value of Indigenous expertise in water management. Specific milestones include the Echuca Declaration; the appointment of Indigenous people to Basin water boards and the Murray–Darling Basin Authority; and explicit recognition—by the Council of Australian Governments and in the 2004 National Water Initiative—of Indigenous interests in water.

First adopted in 2007, the Echuca Declaration reaffirms that each of the Murray and Lower Darling Rivers Indigenous Nations is, and has been since time immemorial, sovereign over its own land and waters, and that people of each Indigenous Nation obtain and maintain their spiritual and cultural identity, life and livelihood from their lands and waters. 'We,' the Declaration states, 'the Indigenous Nations of the Murray and Lower Darling Rivers have never given up our sovereignty over our Country and it is our Country that has always given us everything.'

In 1969, archaeologist Rhys Jones challenged the blinkered but widely held view that Aboriginal people were beholden to the environment, in contrast to European settlers who were seen as 'masters of Nature'. When he published his paper 'Fire-Stick Farming', Indigenous Australians had only recently achieved the right to vote. Australia was still burdened with the false legal doctrine of *terra nullius*.

We now know Indigenous Australians have long used sophisticated strategies to manage the land and the environment. Cultural burning is an example. Also known as 'cool burning', it was an efficient way to manage grasslands and to open tree canopies, so understory plants could germinate, and the biodiversity and carrying capacity of the land would increase.

Indigenous people also actively managed the landscape by collecting and distributing seeds, and by curating the habitats of ground-dwelling prey. They learned how to endure through floods and drought—including by finding and managing water—and they shared their knowledge and expertise across the Basin.

Lake Mungo, a dry lake, is part of the World Heritage–listed Willandra Lakes Regions. The Paakantji (or Barkindji or Barkandji), Ngiampaa and Mutthi Mutthi peoples are joint custodians of this sacred land. The lake is not far from the town of Pooncarie (originally gazetted as 'Pooncaira', the Paakantji word for 'large sandhill') on the Darling River between Wentworth and Menindee, about 760 kilometres west of Sydney.

The Lake Mungo sedimentary layer was formed long before the last ice age. Rainfall from the Great Dividing Range, far to the east, kept the lake full during times of low local rainfall. The lake and its surrounds were rich in biodiversity: many varieties of megafauna roamed there, including giant kangaroos, hairy-nosed wombats and a hippo-like animal called Zygomaturus.

During the ice age, the water level dropped and the watery lake became a salt lake. The salt made the soil alkaline, which helped preserve fossils. A series of 'lunettes' (crescent-shaped dunes) up to 40 metres high and stretching more than 33 kilometres formed the eastern side of the salt lake. These features are known locally as the Walls of China, and are a dream for budding palaeontologists.

Salt-tolerant vegetation helped stabilise and protect the area from erosion—until the arrival of Europeans and their goats, sheep and rabbits. The removal of vegetation, combined with increased visitation, led to ecological damage but also marvellous discoveries, including a set of 20,000-year-old footprints on the shore of the dry lake. Extensive middens were also found, consisting of sandstone tools, fireplaces, and the discarded shells of yabbies and mussels and crayfish.

Estimated to be around 10,000 years old, the middens preserved evidence of how people lived in the area for millennia. They reveal extensive exchange and travel. One stone axe-head found in the dunes is more than 500 years old and was made from stone from Mount Camel, another Basin location more than 300 kilometres away, between Bendigo and Shepparton.

In 1968, Jim Bowler, a geologist and research fellow at the Australian National University, found human skeletons that had been buried in the former lake bed for more than 40,000 years. The skeletons—the oldest ever found in Australia—include Mungo Lady, whose grave is the world's oldest cremated burial site. With this discovery, Bowler said, he was confronted with 'the very presence of humanity itself'.

In 1974, Bowler returned with his team and this time they discovered the complete skeleton of a man, who became known as Mungo Man. The discoveries of Mungo Lady and Mungo Man put Lake Mungo on the world map and rewrote the official history of Australia and its people.

'This research extends far beyond mere academic interest,' Jim Bowler later said. 'Non-indigenous Australians too often have a desperately limited frame of historical reference. The Lake Mungo region provides a record of land and people that we latter-day arrivals have failed to incorporate into our own Australian psyche. We have yet to penetrate the depths of time and cultural treasures revealed by those ancestors of Indigenous Australians.'

—

Farmers Shelley Scoullar and Lachlan Marshall put us in touch with David Crew to see how Indigenous people in the region were using traditional water, grains and land management methods. David is public office holder and manager of the Yarkuwa Indigenous Knowledge Centre in Deniliquin. A warm, generous local willing to share his unique knowledge to nurture consensus leaders and future custodians of the Basin, he established the centre in 2001 with his wife, Jeanette, a Mutthi Mutthi Elder.

When we met David, he told us about what was happening in the river system and the adjacent forests, and how, with unseasonally high water, people in his community could no longer practise traditional learning and traditional ceremonies and methods, because much of their land was underwater. He also explained how sacred stories and deep traditions helped protect the land and biodiversity.

'We have a creation story,' he said, 'about how the river was made. It talks about the rainbow serpent who made the river and had come up and disturbed the crow that had been sleeping underground after a fight with the eaglehawk in Balranald. When the crow was woken by the rainbow serpent he flew up in a rage and broke the rainbow serpent up into pieces and these become our waterholes, and they all have stories attached to them. They're basically refuge holes. Nobody was allowed to fish in the water because that was the refuge for breeding fish. They've all got story sites that stops people going into them.'

David and Jeanette had resolved to share their knowledge across the Basin in order to improve cultural awareness and farming practices.

'When we first took up the opportunity,' David said, 'to work with the Murray Regional Strategy Group, we said they, too, have to know this history. It's important that the people doing productivity on the land understand what they're doing to this land. They're doing their farming work on top of Aboriginal land.

'And while we get some people who say that they only care about money rather than people, we say, well, you're making money where the First Nations people are losing. So how does that sit comfortably that we've got this? My wife actually mentions once when she spoke with the head of SunRice and she talked about the cultural context of the land that they worked on, and he actually admitted that he never even thought about it.'

Tens of thousands of years ago in south-west Victoria, the eruption of Budj Bim (formerly known as Mount Eccles) created an eight-kilometre-wide, 18-kilometre-long lava flow. More than 6,000 years ago, Indigenous Australians built a unique system within the stony remnants of the flow. That system—as important as any other stone archaeology around the world—features complex channels for water management and aquaculture. It has remained in almost continuous operation.

The local Gunditjmara people, including the Gunditj Mirring Traditional Owners Aboriginal Corporation, worked for decades to ensure the site was properly recognised and protected. In the 2000s, a campaign was launched to add Budj Bim to the World Heritage list. For a perspective on the government steps to protect the site, we spoke to Keith Hamilton, former Victorian minister for agriculture and minister for Aboriginal affairs.

'We were determined,' he said, 'to bridge the divide between Indigenous and non-Indigenous peoples...Recognition of the Budj Bim site would challenge the preconception that Indigenous Australians were all nomads, and therefore unable to establish a continuous connection to land. Challenging that idea was critical in achieving Indigenous rights under white law.'

Kate Auty, the former Victorian commissioner for environmental sustainability and the state's inaugural Koori Court magistrate, also took a keen interest in helping to add Budj Bim to the World Heritage list.

'[Gunditjmara Elders] Iris and Charlie Lovett first talked about a reflood of the [Budj Bim] lake in 1982,' Kate said. 'Then Laura Bell wanted it done in the late '80s, and Damein [Laura's son] and the gang talked about it in the early '90s. Then the World Heritage stuff got some government support. I got a walk over when I was commish in 2011, and the brolgas were back!'

In 2019, Budj Bim entered the United Nations World Heritage List—the 20th Australian site to make the list—alongside Uluru, the Daintree, the Great Barrier Reef and Melbourne's Royal Exhibition Building.

'When this went through,' Kate said, 'I simply felt warmed up. Iris [Lovett] would be jumping for joy.'

Through Kate, and the Gunditj Mirring co-op at Heywood, we contacted local Gunditjmara man Damein Bell, CEO of the Gunditj Mirring Traditional Owners Aboriginal Corporation and a key leader of work being done to recognise and protect Indigenous heritage and science.

'We were inscribed by UNESCO into the World Heritage list for Budj Bim's immense and outstanding universal values to human history,' Damein said. 'There was a couple of tense moments, but at the end of the day the gavel came down and we were inscribed.'

The recognition, Damein said, was something to be celebrated, and it was long overdue.

'Budj Bim is 30,000 years old,' he said. 'The site of the world's oldest freshwater aquaculture system. We know that landscape has done droughts before. We know that our ancestors had engineered the landscape in response to drought, you know, 4,600 years ago. Even tidal waves are part of the Budj Bim history. We still need to learn the lessons of hydrology and how water works, the ways that were engineered by our ancestors.

'We just need to use the technology to look at how it worked in the droughts, how it worked in the floods, and how it maintains its integrity as a sustainable landscape that looked after thousands of people for thousands of years. It's right there within these hundred square kilometres.'

The great majority of pre-European stone-built archaeology in Australia relates to water management. Budj Bim is part of a larger system and a larger tradition of water infrastructure. On the Murrumbidgee, too, there is important stone archaeology. And another key site is on the Darling.

The headwaters of the Darling are far to the north of those of the Murray. There are no highish mountains with melting snow to fill the Darling and its tributaries. It fills from rainfall, mostly from alluvial sedimentary floodplains, and from shallow and narrow rivers in Queensland. Its milky colour is from the silt that accumulates on the slow journey west and south. At the junction of the Murray and Darling near Wentworth, the white water from the north blends with the blue and brown water from the east. Before the waters fully merge, the effect is like mixing paint or marbling with ink.

Between Walgett and Bourke in the Darling Riverine Plains—a semi-arid region that is hot and dry for most of the year—are the Baiame's Ngunnhu, also known as the Nonah or Nyemba Fish Traps. At this site, the Ngemba people used their advanced knowledge of river hydrology and fish ecology to harness the natural waterways in order to catch plentiful yabbies, shrimp, crayfish, catfish, black-bream and yellowbelly.

The custodians of the fish traps are the Ngemba Wayilwan (or Wailwan) people. According to a Ngemba Wayilwan Dreaming story, the ancestral being Baiame revealed the design of the traps by throwing his fishing net over the river. His two sons, Booma-ooma-nowi and Ghinda-inda-mui, used the pattern of the net to build the traps from stones. Still visible today is the intricate design of the dry-stone rock weirs and pens—looking remarkably like a net thrown over the river.

CHAPTER 3
THE INLAND SEA

The first Europeans to arrive in Australia looked down on swamps. Swampy land was poor land, they believed, unsuitable for sheep or cattle or broadacre crops such as wheat, barley and canola. The aim of many settlers was to clear the scrub and drain the swamp, to create land fit for farming.

Captain Arthur Phillip had sailed to Australia's east coast with the intention of settling at Botany Bay. But on arrival in 1788, he reportedly judged the bay to be too shallow and exposed, and the environs too swampy. Fortunately, the opening to a much better harbour was just a short distance to the north. Phillip led the First Fleet through the Heads—which would make useful sites for lookouts and artillery—and into the ancestral Country and fishing grounds of the Eora people.

Philip had arrived on Gadigal land (also spelled Cadigal and

Caddiegal). Here, he experienced 'the satisfaction of finding the finest harbour in the world'.

Since those days, swamps have been reappraised. We now know them as brackish lakes, lagoons, marshes, estuaries and 'forested wetlands', all of which are critically important transition zones for natural ecosystems. But the white man was on a quest to change Australia's landscapes forever. Cities would be built, wild rivers 'tamed', land cleared and swamps drained.

Aboriginal and Torres Strait Islander people knew where water was—the rivers, creeks, waterholes and billabongs—and how to get it. According to historian Michael Cathcart, 'Charles Sturt observed that Aborigines who lived on the Murray were sensitive to the subtlest rises and falls of the river. They knew exactly when the river would transform "all the dried up lagoons" into ponds, teeming with birds and fish.' These water skills were highly valued by thirsty white settlers—an impetus for cooperation and conflict.

The iconic 1813 expedition of Gregory Blaxland and William Charles Wentworth across the Blue Mountains was only possible thanks to the knowledge of the Darug people, who knew two routes: one following the Bilpin Ridge (the path chosen by the explorers) and another along Coxs River.

The crossing of the Blue Mountains overcame the physical barrier to the expansion of the Sydney colony, which was hungry for land and resources, especially after the droughts of 1812 and 1813. Bathurst, Australia's first inland European settlement, was founded on 7 May 1815 when Governor Macquarie raised the flag at the terminus of Coxs Road.

European explorers were convinced they would find in the centre of Australia a sea, comparable to the Mediterranean or the Caspian, or at least a major river cutting through the interior, like those in

Africa and North America, or perhaps a major north–south strait, separating the eastern and western halves of the continent.

In his 1814 work *A Voyage to Terra Australis*, English mariner Matthew Flinders recounted his instructions: 'in case you should discover any creek or opening likely to lead to an inland sea or strait you are at liberty either to examine it or not, as you shall judge it most expedient, until a more favourable opportunity shall enable you so to do'. In 1798, Sir Joseph Banks nursed the hypothesis of a great river. 'It is impossible,' he wrote, 'to conceive that such a body of land, as large as all Europe, does not produce vast rivers, capable of being navigated into the heart of the interior.'

In 1817, New South Wales surveyor general John Oxley and his deputy George Evans left Sydney and arrived in Bathurst before setting off to explore the Lachlan River. They came across marshy country, and Oxley's thesis was that the Lachlan flowed into a series of unamenable swamps. He was 'confident we were in the immediate vicinity of an inland sea, most probably a shoal one'.

If the expeditioners had pressed on for two more days, they would have reached the Murrumbidgee River. But the team became stuck so they turned northwards to follow the Macquarie River back to Bathurst. The expedition up the Macquarie River again came across swampy land, which became known as the Macquarie Marshes. On the explorers' return to Sydney, they shared their impressions and assumptions, and the legend of the inland sea grew.

In 1840, Edward John Eyre did come upon a gigantic lake in central Australia. It covered 9,500 square kilometres—but the lake, about 700 kilometres north of Adelaide, was mostly dry. The lake is now called Kati Thanda–Lake Eyre. About four times in every century, it fills with water from wet season rains in Queensland, which travel through a network of rivers from near Townsville and Mount Isa to the Lake Eyre basin.

The average depth of the full lake is about four metres. The last time it was full was in 1974, but it was partially filled from 2009 to 2012 and again in 2019. In those periods it became a biodiversity wonderland. Pelicans, gulls, banded stilts, sandpipers, terns and native fish including bony bream, golden perch and hardyhead—keenly watched by the Lake Eyre Yacht Club.

In 1844, explorer Charles Sturt left Adelaide with a team of 17 men and a Noah's ark of 11 horses, 30 bullocks and 200 sheep. They even brought a whaleboat. Their goal was to find the true inland sea, perhaps a remnant of the Great Flood. Almost a year later, after battling disease and dehydration and malnutrition, Sturt returned to South Australia. He had had the right idea, but he was about 120 million years too late. During the Cretaceous period, a great sea did cover about a quarter of the great southern land.

Count Paul Strzelecki was the last of the great modern explorers of the Murray. In 1840, he journeyed to the neighbourhood of the river's source. Having struck the Murray near Corryong, he travelled upriver to the foot of Mount Kosciuszko. He marvelled at the landscape, which had 'attractions for the explorer of no ordinary kind'. '[E]very feature of that division,' he wrote, 'seems to bear the stamp of foreign grandeur...the Australian Alps, with its stupendous peaks and domes; and in front the beautiful valley which the Murray so bountifully waters...'

On 15 February 1840, Strzelecki ascended mainland Australia's highest peak: 'elevated above all the heights...the snowy and craggy cone of Mount Kosciuszko is seen cresting the Australian Alps in all the sublimity of mountain scenery'.

In his *Physical Description of New South Wales and Van Diemen's Land*, Strzelecki described his awestruck reaction to the Murray's source. '[L]ooking from the very verge of the cone downwards almost perpendicularly, the eye plunges into a fearful gorge 3,000 feet deep, in the bed of which the sources of the Murray gather their contents and roll their united waters to the west. To follow the course of that river from this gorge into its farther windings, is to pass from the sublime to the beautiful.'

CHAPTER 4
THE TRAGEDY OF THE COMMONS

In 1833, the British economist William Forster Lloyd coined the idea of 'the tragedy of the commons'. Without careful management and collective agreement, Lloyd posited, people would likely misuse and overuse shared resources such as grazing land, waterways and public open space. The unchecked actions of individuals would harm the collective good.

American ecologist Garrett Hardin popularised the phrase and the concept. Like Lloyd, he feared that, as farmers raised more livestock on shared land, their own profits would increase for a time but so would the shared costs. The tragedy in this parable is that eventually no farmer can use the land due to overconsumption. Everyone loses.

The tragedy of the commons has become an urgent issue in unexpected places, such as the world's oceans, coastal land and even the office fridge. (At the risk of incurring the Wrath of Trekkies, it reminds us of the Spockian wisdom: 'Logic clearly dictates that the needs of the many outweigh the needs of the few'.)

Environmental scholars and policymakers today view the tragedy of the commons through multiple lenses, including intergenerational equity. How will future generations fare if we mismanage common goods such as land, soil and air? Water, too, is a classic common good and the Murray–Darling Basin is a classic example of the tragedy of the commons, one made even more difficult by Australia's political geography.

The first European settlers had come from a wet country. Flush rivers and streams nourished English lakes and Scottish lochs. In Australia, the white drover-explorers who followed the tracks of Blaxland, Wentworth and Strzelecki saw the potential of rivers such as the Murray and the Murrumbidgee. With the opening of ways into inland Australia, squatters quickly claimed vast lands along those watercourses. By the 1840s, squatters and settlers had picked the best grass flats for cattle camps and sheep stations.

The spread of Europeans was usually a calamity for Indigenous people. Just one example: in 1841, the first chief commissioner of police in South Australia, Major O'Halloran, with Matthew Moorhouse, who had been given the title of the protector of Aborigines, rode out from Adelaide with a bushmen's regiment of 68 volunteers. The result was the Battle of the Rufus, a deadly and one-sided conflict that pitted spears against rifles. Fought on the shores of Lake Victoria, it claimed the lives of at least 50 Indigenous people.

In the early 1880s, a severe drought devastated farms in the north and north-west of Victoria. The government teamed up with private railways to deliver emergency water on specially configured 'water trains'. The drought spurred the creation of water and irrigation trusts and the construction of reservoirs and weirs and channels.

Alfred Deakin was Victoria's chief minister and the chief architect of the colony's water policy. Influenced by Victorian MP and irrigation advocate Hugh McColl, Deakin viewed the harnessing of water as a vital instrument of national progress and economic development, especially in areas of precarious rainfall. (An 1883 caricature depicts McColl with an 'Unprecedented case of Hydrocephalus'.) According to Deakin biographer Judith Brett, Deakin's heroes were 'the men of the trowel rather than the sword'.

On 23 December 1884, the Victorian government, responding to McColl, announced a Royal Commission on Water Supply and Irrigation under the chairmanship of Deakin as solicitor-general and commissioner for public works and water supply. It became known as the Deakin royal commission and it would run from 1884 to 1887.

Within days of the announcement of the royal commission, Deakin embarked on an investigative trip to the United States. His family would join him, along with journalists John Dow of the *Age* and E. S. Cunningham of the *Argus*, plus J. D. Derry, a state engineer with experience of irrigation in India. The whirlwind mission would take in Arizona, California, Colorado, Kansas and Nevada. The Australians would inspect irrigation works that watered orangeries, vineyards and orchards, and they would meet the engineers and entrepreneurs who orchestrated the works.

The trip convinced Deakin that Victoria was just as suitable for irrigation as California. It also convinced him that 'the government needed to take control of water rights before vested interests became too deeply involved'. To members of the water supply royal commission, Deakin addressed a 15 June 1885 progress report, titled *Irrigation in Western America, so far as it has relation to the circumstances of Victoria*.

> A recognition of the danger of allowing water to be monopolized without regard to the land has led a commission appointed to inquire into Californian irrigation to declare that, 'as a matter of public policy, it is desirable that the land and water be joined never to be cut asunder; that the farmers would enjoy in perpetuity the use of the water necessary for the irrigation of their respective lands; that, when the land is sold, the right to water shall also be sold with it, and that neither shall be sold separately.'

Deakin quoted Major John Wesley Powell, director of the US Geographical and Geological Survey Department, who, 'in his careful draft of a land system adapted to the arid region, most emphatically recommends that "The right to use water should inhere in the land to be irrigated, and water-rights should go with land titles"'.

'In Colorado,' Deakin further wrote, 'the feeling has gone so far that a proposal has been made in the Legislature to compel all canal owners to supply any persons with water, which they are not themselves using, at fixed rates; but as this would simply mean transferring to landowners who had invested nothing in canals part of the profit to be made by those who had so invested, the proposal was not entertained.'

A major step forward on the subject, the report concluded:

> 1. The state must exercise supreme control of ownership over all the rivers, lakes, streams, and sources of water supply, except springs on private lands.

2. The state should dispose of the water to irrigators, and should be guided in doing so by its own qualified personnel, in order to encourage the greatest possible utilisation of water on the largest possible area.
3. A standard scale of water measurement should be established for use in all water transactions.
4. Local 'water masters' should be appointed to supervise distributions and settle disputes; the duties of these officials should be organised by a central office to 'guarantee the preservation of watercourses and other sources of supply'.
5. Where satisfactory proof of need was given and due compensation paid, power of easement over private lands should be enjoyed by irrigators.
6. The state should provide as much information as possible about the natural capacities of its territory for irrigation.

These and other findings were the basis for Deakin's seminal *Irrigation Act 1886 (Victoria).*

Breaking with English and American legal precedents, the new act established that no private individual could own a river or control the use of its water. The 'use and flow' of water resources were vested in the Crown. 'Under Deakin's scheme,' wrote Michael Cathcart in *The Water Dreamers*, 'the water, in effect, belonged to the river—and the irrigators had an entitlement to use an allocated amount on their farms, for which they paid a small annual fee.' This allowed the colony to establish centralised bureaucratic systems for allocating water rights.

The underlying principle was emphatic. In Australia, water was so precious that no individual should profit from or monopolise its control. Water policy should be decided not by money men and water barons, but by the democratically elected parliament. This principle was so compelling that other Australian colonies soon introduced similar provisions.

Starting with an orange orchard in the foothills of the Sierra Madre, the Canadian brothers George and William Chaffey refined a money-making scheme that involved colonisation through irrigation. They would buy land, irrigate it, develop it, then sell it in blocks, use the money to fund development, then finally reap profits through the appreciation in the land's value.

In 1885 in Los Angeles, Deakin had met George Chaffey, who with William had been in the process of selling parcels of 'Ontario' (sometimes written as 'Ontaria'), a Californian estate of 10,000 irrigated acres. The following year, Stephen Cureton, an agent of the Chaffey brothers, brought George to see Deakin in Melbourne. Cureton has been described as 'a fourth-rate pressman of poor education, whose sole accomplishment was a plausible tongue'. In America, he'd regaled the Chaffey brothers about the boundless plains of plenty along the Murray. Now in Australia, he spruiked to Deakin the prospect of Chaffey investment on the southern continent.

In February 1886, George Chaffey arrived in Australia full of enthusiasm. Judith Brett noted that 'Deakin met with him regularly during that year as they negotiated an agreement'. Convinced that this was his destiny, to be nation-building with Deakin, George cabled William, who sold the brothers' remaining interests in California and set sail for Australia.

Deakin had found California fascinating and informative, though with fundamental differences from Victoria. California did not need the huge storages that would be required to 'drought-proof' Australia, for example. California also had extensive access to cheap labour, in the form of Mexican workers, who were paid starvation wages—a fact unacceptable to Deakin's egalitarianism.

Americans saw less of a role for the state in the provision of essential utilities. As a consequence, private interests owned most of America's irrigation infrastructure. Other differences between California and Australia were cultural. Deakin had been 'repelled by the self-interest of Americans' frontier individualism', and especially by how irrigators were ever ready to engage their neighbours in acrimonious litigation over riparian rights.

In Australia, the Chaffey brothers felt these contrasts in reverse. American investments were made in a quickfire, 'just do it' way. In Australia, things moved less rapidly and there were more regulations about what could and could not be done. An act of parliament would need to be passed before the Chaffey brothers could plant an orange tree on their preferred acreage.

The Waterworks Construction Encouragement Bill was introduced to the Victorian parliament in 1886 with the goal of establishing a thriving irrigation settlement in the colony's north-west corner. That venture drew criticism as a 'Yankee land grab'. To reassure the critics, the government put the scheme out to public tender. A single bid was received—from the Chaffey brothers—and the bill finally passed in December 1886.

To 'secure the application of private capital to the construction of irrigation works, and the establishment of a system of instruction in practical irrigation', the Chaffey brothers were granted an allotment of 50,000 acres, with the right to purchase the adjoining 200,000 acres at £1 per acre. California's Ontario became the template for Victoria's Mildura.

'George Chaffey was no fool,' said Australian journalist, travel writer and novelist Ernestine Hill in her excellent book *Water Into Gold*, first published in 1937. 'Here were the soils of California, the suns of Spain, and the third ingredient—water: 1,088,000,000,000 gallons of it—rolling past every year.'

A century after the arrival of the First Fleet, a speculative land and housing boom was in full swing in Australia, and especially in Victoria. Government debt was mounting. On 30 October 1890, Victoria's Gillies–Deakin government was defeated. 'The main issue,' Brett explained in her Deakin biography, '[was] financial extravagance and the mountain of public debt. The agreement with the Chaffey brothers came in for special attack as an example of gross maladministration. Deakin vigorously defended it and talked up the prospects for Mildura.'

Newspaper proprietor David Syme persuaded Deakin to visit India. Just as he'd done in America, Deakin made a flying tour of the country, inspecting the major irrigation works of the Ganges and the Punjab, which secured the country against famine. 'He believed he was the first white man, other than officials, to travel up the Sirhind Canal in Lahore, on a boat pulled by natives, sleeping on a hinged ledge with his boots for a pillow.'

In 1891, Victoria's economic boom came to an abrupt end. Soon banks and building societies were closing. Bankruptcies and unemployment surged. Deakin spent the 1890s as a backbencher in the Victorian parliament. He was 'deeply disillusioned by the financial and banking scandals that engulfed the colony'. He'd been on building society boards and felt implicated in the loss of depositors' savings, including his own and his father's.

The effects of the Depression touched the whole state—even distant Mildura. Demand for fruit fell, credit tightened, and irrigation channels leaked. In 1893, the Chaffey brothers put their company into liquidation. Two years later they filed for bankruptcy amid allegations that they were a 'gang of swindlers', taking Australians for a ride. Thanks to this episode, an aura of corruption and fraud would thereafter surround ambitious irrigation schemes in the Basin. Having embraced and then talked up the Chaffeys' scheme, Alfred Deakin now had to defend his own judgment and integrity. He did so in newspapers and then at the royal commission that inquired into the affair. As Brett wrote:

> He argued in his defence that problems had only emerged after 1890 when he was no longer in government, and that the Chaffeys had overstretched themselves by taking on a scheme at Renmark for the South Australian government. Defending the government's initial decision, he remained optimistic about the future of the Mildura settlement, particularly if a railway could be built to get the settlers' fruit to market.

Deakin reportedly considered leaving politics, but decided to stay on and help craft the Constitution. He wanted to ensure his egalitarian principles were reflected in Australia's founding document. That way, the nation could avoid the tragedy of the commons.

CHAPTER 5
COOPERATION

The first attempts by Europeans to control Australia's inland hydrology were clumsy and temporary. The colonists largely ignored Indigenous people's deep and sophisticated connections to land and water and Country. Indigenous boundaries aligned well with how water moved in the landscape. But the political and economic structure of modern Australia cut across catchment boundaries. The die was cast for more and more fighting over less and less water.

When the colonies came together to draft a potential national Constitution, Australia was in the grip of the Federation Drought (1895–1902). The sharing of scarce water was among the most hotly contested issues. Farmers and irrigators pushed for ongoing rights to and control of water. To satisfy them, the Constitution stated: 'The Commonwealth shall not, by any law or regulation of trade or commerce, abridge the right of a State or of the residents therein to the reasonable use of the waters of rivers for conservation or irrigation.'

As Commissioner Bret Walker later wrote in the report of a South Australian royal commission on the Murray–Darling Basin, 'In the absence of utterly unrealistic change to our Commonwealth Constitution by referendum, this very Australian framework of governance will remain, *faute de mieux*. The cheerful term for it is "co-operative federalism". A grimmer view would see it as a cockpit for interstate rivalrous self-interests.'

In the early years of Federation, the country had three main parties: Protectionists, Free Traders and Labor. At that critical moment in Australia's history, there was even more political turbulence than we have today. Yet cross-party support on the Murray–Darling survived. In one of the earliest examples of 'cooperative federalism', the River Murray Water Agreement 1914 was established. The Commonwealth, New South Wales, Victoria and South Australia all signed up to the agreement, which was informed by the 1902 Interstate Royal Commission on the River Murray, and which set out the shares of water available to each state.

The River Murray Water Agreement was ratified in 1915 when the Basin states and the Commonwealth government passed enabling legislation. In 1917, the River Murray Commission came into being under the intergovernmental agreement. The commission regulated the construction of structures such as weirs and locks along the Murray.

From around 1917 to 1990, engineers reigned supreme in Australian water management. This long period was an era of big projects and grand visions. After World War I, locks, weirs, dams, barrages, lakes and generators were built along the Murray. Rivers were tamed and harnessed for agricultural productivity and economic output.

Weir and Lock 1 at Blanchetown on the Murray were the first to be completed, in 1922. Other weirs and locks were progressively constructed. Every major town on the Murray and Darling rivers would have a weir to store water for consumptive use and to regulate river flows. Yarrawonga Weir, completed in 1939, was the only weir on the Murray River below the Hume Dam that did not have a lock.

Over the next decades, much larger dams were also built. Water was taken from the system, to the benefit of irrigators and some communities, but with little care for the short- or long-term impacts on the system as a whole, or the Basin environment.

About 110 kilometres south-east of Broken Hill, the Menindee Lakes were once a series of natural depressions along the Darling River. During floods they filled and at other times they were mostly dry. But in the 1950s, a major program of works turned the depressions into a big gated dam. From that point on, the lakes could be filled to a total volume of 2 billion litres. When full, they have a surface area of about 50 square kilometres.

In 1963, the lakes and their water storage were the focus of an important instance of cooperative federalism: a new River Murray Waters Agreement between the Commonwealth, New South Wales, Victoria and South Australia.

These four governments had worked together over decades to build the locks, weirs and dams, and to share the costs and benefits. The intergovernmental collaboration was more or less consistent, and the cost-sharing was more or less equitable. But there was one notable exception.

In the early 1960s, the governments proposed the Chowilla Dam as a large water storage on the Murray River. The dam wall was to be built in South Australia, but the reservoir itself would stretch upstream into Victoria and New South Wales. In concert with the River Murray Commission, the four governments agreed to share the project's costs—and water.

The proposal depended on the Commonwealth extending a loan for the New South Wales part, in return for the water from the Menindee Lakes. As with most water infrastructure projects, the expected cost blew out, and concerns over the environmental and hydrological impacts grew. The particular concerns with this large and shallow reservoir, in such a hot and dry region, were many. What would be the losses from evaporation? Would the dam exacerbate salinity and cause excessive irrigation?

Like water in a shallow dam, cross-state and cross-party support for the Chowilla Dam evaporated. The dam's original proponent, South Australian premier Tom Playford of the Liberal and Country League (LCL), only remained in power in 1962 with the support of independent Tom Stott, whose electorate included the proposed dam. When the South Australian water commissioner (in defiance of his premier) supported the building of the Dartmouth Dam in Victoria rather than the Chowilla Dam, Stott crossed the floor and the government fell. By the time the LCL came back into power in 1968, its new leader, Steele Hall, had changed his position to favour the Dartmouth option. Victoria also walked away from Chowilla.

Liberal Sir William McMahon ('Billy Big Ears') was prime minister from 1971 until 1972, after forcing the resignation of John Gorton. Under McMahon's leadership, the Commonwealth used its financial power to pressure the states to adopt a whole-of-basin perspective.

Eventually the deadlock was broken and the 4-billion-litre capacity Dartmouth Dam was built on the Mitta Mitta River, a tributary of the Murray, in north-east Victoria. Agreement, it seems, was easier to achieve in the Murray's mountainous upper reaches than in its arid flatlands.

In 1981, Liberal prime minister Malcolm Fraser met with the Basin premiers to agree on extending the role of the River Murray Commission to cover water quality and other matters. This led to the first Murray–Darling Basin Agreement in 1982 and the beginning of total catchment management. The *Murray–Darling Basin Act 1983* was ratified and passed by four parliaments.

Basin state governments established irrigation areas at St George in Queensland, in the New South Wales Murray and Murrumbidgee valleys, and in Victoria's Goulburn–Murray and Sunraysia regions. 'These purpose-built irrigation areas,' Griffith businessman Paul Pierotti said, 'are completely and solely reliant on the water and the industries that exist around them.'

The regions became Australia's foodbowl. Individual farms were scaled to suit family farmers, and they were designed for regular irrigation. Farmers got to know their land, said farmer Shelley Scoullar, 'the strengths and weakness of different paddocks, where certain weeds were likely to pop up, which paddocks used water the most efficiently, spots where banks were leaky'. Water was the key to turning unproductive land into productive land, and a marginal crop into a good crop.

'I loved watching the crops develop and change over their growing season,' Shelley said. 'To see a deep green rice crop swaying in the breeze was so soothing, the bright yellow of a good canola crop and the purple flowers of a field pea crop is something special and made you feel one with nature. It is hard to describe the feeling of working with nature to give a crop the best opportunity to reach its potential so that you can provide food for others. It gives you strength and soul.'

'These farms in the Riverina,' farmer and barrister Robyn Wheeler said, 'most of them are family farms, and they do a really, really good job at growing food and fibre for the country. They're very efficient. MIL's [Murray Irrigation Limited] irrigation system is one of the best in the world. It recycles water. Properties on the MIL system have turkey-nest dams. Water is not allowed to run off. We're not allowed to have salt build up. We recycle the water. We've got new gates being developed all the time. Water gets on and off quickly. I mean we use it, and reuse it, and it's amazing country if you've ever been in the Riverina. It responds so well because it's such fabulous soil. Just the smallest amount of water and off it goes. It's quite something.'

The foodbowl farms were developed 'to feed the nation', Robyn said. 'So every property had a water entitlement attached to it. And it was not separable from the land. So if you sold your land, your water went with it, part and parcel.'

Rules and practices differed from place to place, but broadly speaking, blocks were allocated rights to sufficient water for irrigation. The combination of water and land could have remarkable effects.

'I can really distinctly remember,' Shelley Scoullar said, 'when that first bit of water came down the channel and into the paddock. I just knew this is what I was put on earth to do. To grow rice. The frogs just come alive and I would find any excuse to go check the rice. I'd leave the dishes for my husband to do and I'd just find any reason to be out there. I loved it. I wanted to be efficient. I wanted to get better. I just loved, loved, loved it.'

PART II: MAKING THE MARKET

CHAPTER 6
WATER FOR BEER

In line with Alfred Deakin's legacy, Australian governments were for a long time wary of large corporate entities and foreign owners monopolising irrigation water and rural land. To minimise 'the corporatisation of agricultural land and water rights', governments limited farm sizes in public irrigation developments to the estimated needs of a family farm; they limited the proportion of any single land parcel that could be irrigated; and they tied water rights to the land on which the water was to be used.

When you bought farming land, therefore, you also acquired the water entitlements that went with that land. But the idea of selling the water entitlement, or otherwise trading the associated water, had early precedents across the Basin. There are reports of informal water trading during World War II, for example, so irrigators had enough water to endure the 'water famine' when drought conditions prevailed between 1937 and 1945.

Through short-term and unofficial trades in that era, 'a farmer would pay some money to his neighbour and then simply ask the water bailiff to "send Joe's water down to me for the next two weeks"'. That informal tradition continued well into the 1980s and 1990s. 'Growing up,' farmer Shelley Scoullar told us, 'if you needed a few more megs of water to finish off watering your wheat or whatever, you gave your neighbour a slab of beer—for a few megs of water. By the time I returned to the land in 2006, that had obviously changed quite significantly.'

Murray Irrigation's George Warne told a similar story when he spoke to the ABC's *Four Corners* in 2003: 'If you said to someone ten years ago, $300 a megalitre, he would laugh at you, there were hundreds if not thousands of megalitres being traded for something as simple as one slab of beer.'

Beginning in the 1960s, irrigation groups and Basin states had conducted various kinds of more formal water trading experiments and market trials, first within irrigation districts and then within and between catchments and valleys. The limited trials and experiments were mostly seen as necessary responses to isolated and temporary water shortages. During the drought of 1966–67, for example, temporary transfers of water rights were permitted in New South Wales and Victoria. In the 1972–73 drought, New South Wales again allowed temporary trades. In Victoria, limited trading was permitted from 1982 to 1983.

After this faltering start, widespread temporary water trading began in 1983 in New South Wales and South Australia, and in 1986 in Victoria. New South Wales allowed permanent water transfers from 1986. In Queensland, trade in annual allocations began in 1987–88 in defined irrigation areas on an experimental basis. The areas included the Border Rivers region, St George and Bundaberg, and the Upper Condamine and Emerald schemes. To prevent windfall gains, only irrigators who could prove a history of use could trade water allocations. Amid pressure from irrigators, a statutory basis for temporary trade was introduced in the *Water Resources Act 1989 (Qld)*, but no provision was made for trade in permanent rights.

For governments and policymakers, more extensive water trading among irrigators was seen as a way to manage a scarce resource, and to spur improvements in water management and on-farm water efficiency. The expectation was that water rights would flow to where they were most valued: a given megalitre of water would go to the farmer who could use it most efficiently and productively to generate the highest economic return.

According to a 2006 review of water markets by the Productivity Commission, 'Trading can reveal the opportunity cost of water to the community—that is, the benefit forgone by not using water in its best alternative use—and, through mutually beneficial trades, facilitate the movement of water to regions and for uses where it is most highly valued.'

Trading was intended to create a strong financial incentive for farmers and irrigators to use water efficiently. In theory, the most water-efficient farms would thrive, and the system as a whole would maximise productivity for a given supply of water. For the Murray–Darling Basin, well-functioning markets offered a form of insurance for managing climate risk and unexpected seasonal eventualities.

In the 1980s and early 1990s, libertarian economic ideas were rampant in the capitalist West. This was the era of Ronald Reagan, Margaret Thatcher and the Nobel Prize–winning economist Milton

Friedman. Politicians and policymakers put into practice an extreme version of Adam Smith's 'invisible hand', which was supposed to be how free markets would lead, almost magically, to a socially beneficial conclusion. In Friedman's words, 'The government solution to a problem is usually as bad as the problem.'

Friedman's ideas and neoclassical economics were readymade for the post-Soviet era of capitalist confidence and triumphalism. The concepts would take root around the world, including in the offices of Australian policymakers and economic regulators. There was limited knowledge about how precisely to implement the ideas in a way that protected communities and the public interest, and did not create perverse consequences. But the market concepts came to dominate public policy circles anyway, and they intruded into other places, too, including the management of public utilities and environmental assets.

Against that backdrop, the water market was a natural place to experiment with free-market principles. A 1984 meeting of the Australian Water Resources Council and the Australian Agricultural Economics Society heard how 'market forces' would be key to Australia's continued economic growth and optimising the allocation of resources.

'The primary mechanism,' the meeting concluded, 'for achieving this allocation is to expose production processes to market forces, with inputs and outputs valued, as far as practicable, at their economic cost. The opening of a market in water rights/entitlements is seen by many as an important initiative in addressing water allocation problems and wider issues of structure adjustment in rural industry.'

In 1992, Jeff Kennett became premier in Victoria. Influenced by Thatcher-era privatisations in the UK, the Kennett government embarked on its own ambitious program of asset sales, corporatisations and flotations. Merchant bankers and consultants who had worked in economic reform in the UK and New Zealand now arrived in Melbourne to steer Victoria's multibillion-dollar energy privatisations. When those transactions were completed, many of the bankers and consultants stayed on to drive the design of Australia's water markets.

These advisers, and the policymakers they advised, had little confidence in government-driven water development. Instead, the free marketeers embraced the idea that the decentralised decisions of market participants would achieve a more efficient water allocation than central planners ever could. Once let loose, the advisers hoped, 'market forces' would turn water trading into a vital economic engine.

In 1994, the Council of Australian Governments (COAG) agreed to a Water Reform Framework, which was a step towards much more extensive water trading. The framework set in train processes to clarify property rights in water, and to separate those rights from land. The following year, the Water Reform Framework formally became part of the National Competition Policy, which meant the states had to implement it, under a regime that rewarded or punished their progress.

The goals of the Water Reform Framework came to fruition a decade later, in the middle of the 15-year Millennium Drought. In 2004, COAG established the National Water Initiative (NWI), which made water markets a primary tool of water management and the allocation of water in the Basin. Heralded as 'Australia's blueprint for water reform', the NWI was an attempt to improve water pricing, expand trade, introduce registers and water accounting, and require preparation of water plans. Under the NWI, the states would separate water from land—finally and irrevocably—and remove barriers to trade in the separated and 'unbundled' rights.

That innocuous-sounding term—'unbundling'—would soon be a water buzzword. In the seemingly simple concept of water for irrigation, it turns out there was significant hidden complexity. Under the NWI, water rights were 'unbundled' into four separate components:

- **a permanent water access entitlement:** a right to extract a share of available water in perpetuity.
- **a temporary water allocation:** a 'unit of opportunity', usually expressed as a seasonal volume of water available under an entitlement, distributed periodically and tracked in a 'water allocation account' or 'allocation bank account'.
- **water delivery:** the right to have an allocation of water delivered to a specific offtake location (typically using the infrastructure of the relevant irrigation district) or to obtain water from a particular location.
- **water use:** permission to actually use a delivered allocation, within specified conditions and in light of obligations to third parties.

The states welcomed the idea of unbundling. According to the Victorian government's 2004 *Securing Our Water Future Together* policy paper:

> The 'unbundling' of entitlements into their components will create benefits for irrigators. It will: make trade easier, by separating tradeable elements from other

> elements; reduce borrowing costs, by providing for mortgages directly over water; assist leasing, by recording the shares of delivery capacity of both people leasing out and people leasing in; enable a brokering body to offer products tailored to irrigation demand; and make it easier for irrigators to adjust either the reliability of their water supplies, or the timeliness of having water delivered, to suit their individual enterprises.

Banks, too, urged the Basin governments to go ahead with water trading.

In 2007, a new Commonwealth Water Act gave legislative backing to unbundling. According to farmer and barrister Robyn Wheeler, 'The act was designed to get the environmental vote and save the Howard government. It didn't do that, but nevertheless the Water Act came in and separated water from land.' The act removed trade barriers and introduced practical concepts such as 'carryover', which allowed farmers who had unused water in their allocation account to use that water in subsequent seasons.

To fully implement the NWI, the Basin states had to amend their own water legislation. In July 2007, despite Alfred Deakin's advice from more than a century earlier, Victoria took the elemental step of legally separating water from land, and unbundling water rights—into water shares, a delivery share and a water use licence. To inform farmers about the changes, Victorian water authority Goulburn Murray Water sent out 15,000 information packs and held 35 information sessions in that month alone. The authority's CEO, Russell Cooper, said the changes would provide greater flexibility for irrigators.

'At the moment,' Cooper told reporters at the time, 'if an irrigator wants to sell some water, they're obliged by the current legislation to take out an advertisement in the paper...before they can start the process. After July that won't be necessary, we'll have what's termed a water register and on that water register there'll be the current entitlement of all the owners logged there.'

In 2009, South Australia followed suit, amending its *Natural Resources Management Act 2004* and enforcing the separation of water from land, with unbundling in the River Murray Prescribed Watercourse and the Southern Basins and Musgrave Prescribed Wells Area. (The licensing arrangements for other South Australian water allocation plans continued to operate as 'bundled' water licensing regimes. This meant each licence holder's rights for taking and use of water were authorised in one licence.)

New South Wales had commenced its unbundling in 2004, under the *Water Management Act 2000*, but took longer than the other states to complete it. The unbundling process was still underway in 2010, and would extend well into the next decade. A report prepared for the state's Independent Pricing and Regulatory Tribunal in August 2018 found 'rights to water in NSW have now, for the most part, been unbundled from land'.

Separating water from land and breaking water rights into components were two critical turning points in how Australia managed one of its most important natural assets.

CHAPTER 7
THE EXPERIMENT

Gradually, a complex set of water market trading rules were developed to allow water rights to be exchanged and—it was hoped—to allow for water to flow to the right places. The experiment was as grand as it was risky. Successfully applying new market concepts and economic theories to a complicated river system would depend on a triumph of hope over experience.

The physical structures of the river system and irrigation systems affected how water could be traded. The southern part of the Murray–Darling Basin, for example, is more connected: there are more natural and artificial connections between farms and irrigation districts. There is also more water, the valleys are closer together, and trading rules and arrangements developed earlier in the south than in the north. For these and other reasons, including political and historical ones, there was more water trading in the south.

As ultimately conceived, multiple unbundled parts of water rights would be traded. In the allocation market, farmers would buy and sell temporary water rights. There would also be trading in medium-term water leases. And in the entitlement market, farmers would make permanent transfers of water rights. Once a permanent right was sold, the seller would no longer have an economic or legal interest in it.

Rather than have a central exchange or trading platform, the design of the water market saw multiple public and private online exchanges where farmers would trade the different types of water rights. Water brokers would help connect willing buyers and sellers. The Australian Competition and Consumer Commission (ACCC) would monitor the market, and it and other government bodies would help make sure that everything worked and the experiment went smoothly.

To give 'market forces' the greatest chance to allocate water rights efficiently, Basin governments and policymakers sought to have as few rules as possible, including rules about inter-regional trade. But farming groups and National Party politicians, among others, were wary of a fully free market in water. They expressed concerns about the potential for stranded assets, for example, and social harm in regions that suffered an exodus of water rights. Such concerns led Basin governments, irrigation trusts and water boards to limit how much water—and particularly how much permanent water—could move out of a given region or irrigation district.

In South Australia, for example, the board of the Central Irrigation Trust allowed only 2 per cent of an area's water entitlement to be transferred out each year. In 1994, when inter-district trading became possible, the Victorian government imposed a 2 per cent limit on the overall trading of water access entitlement out of a northern Victorian irrigation district in any given season. This was increased to 4 per cent in July 2006. Some New South Wales irrigation districts prohibited outward entitlement trading altogether.

In the 1970s and '80s, there was a battle of ideologies in the economics faculties of Australia's top universities. The victors of this battle were influenced by American economists from Chicago and San Francisco, and were extremely sceptical of government intervention in markets. They trained a generation of like-minded economists, who went on to populate Australia's main economic agencies.

In 1998, the former Industry Commission was given the oddly Orwellian or Stakhanovite name 'Productivity Commission', universally abbreviated as PC. The shift in name marked a shift in philosophy, away from 'industries assistance' and towards free-market, laissez faire, neoclassical economic precepts. With the similarly oriented ACCC, the PC became a prominent advocate for economic reform.

When it came to irrigation, the PC and ACCC shared a vision as clear as Buckland River meltwater. The water market, they consistently argued, should be as unrestricted as possible. That way, the discipline of the market would be strongest, and water would be in the best hands. Restrictions on water trade, they argued, would reduce the effectiveness of the market as a whole, and ran counter to the National Competition Policy, a fact that was used from time to time to convince and coerce Basin governments.

When the National Water Initiative was established, it provided for an interim annual limit of 4 per cent on the volume of permanent trades out of water irrigation areas. The Basin governments agreed to monitor the limit, review it in 2009, and if appropriate lift it further. By 2010, the ACCC was pushing strongly for removal of the 4 per cent limit. According to the ACCC, the segregation of water markets inside and outside irrigation districts impaired the efficient reallocation of water, to the detriment both of sellers inside the districts subject to the limit, and potential purchasers outside:

> The ACCC notes that New South Wales is considering a new implementation of the 4 per cent limit. The ACCC remains of the view that the 4 per cent limit has a number of negative effects, including: preventing water moving to its highest value use; artificially segmenting the water market; leading to efficiency losses, especially long-term dynamic efficiency losses, and preventing the efficient operation of water markets; [and] preventing irrigators experiencing financial distress from realising the value of their water access rights and restricting purchases of water for the environment.

Banks, too, continued to push for a free market. In January 2010, a submission to the ACCC from the Australian Bankers' Association (ABA) expressed support for the propositions that 'there should be no restrictions on trade that relate to the use of water'; 'there should be free trade in and out of the Murray–Darling Basin'; 'trading rules should provide that there are no restrictions on trade specific to water carried over'; and 'trade between regulated system trading zones should only be restricted based on physical constraints, environmental constraints, or hydrological issues'. The ABA might as well have said 'Let it rip'.

There were supportive voices, too, among international consultants and foreign experts, and even among local irrigators. Danny O'Brien, the then CEO of the National Irrigators' Council, wrote in an opinion piece in the *Weekly Times*:

> There are now two reasons why the cap has become a hindrance to Victorian irrigators, not a help. The first is the ability of irrigators to manage their own businesses...Given current commodity prices, it's no surprise many irrigators are trying to sell water. Not only is it unfair to stop them doing so with the 4 per cent cap, it's counterproductive if they go broke in the meantime.

But the states, especially New South Wales and Victoria, defended the limit. As part of a circuit-breaker negotiated with the Commonwealth government in 2009, the Victorian government agreed to a package of exemptions to the 4 per cent limit. The remaining limits, however, still restricted interstate trade, and that fact was the subject of a 2009 High Court case between South Australia and Victoria.

The Constitution, South Australia argued, allowed free trade and commerce between the states, but there was not free trade in water. In particular, South Australia argued, Victoria's water trading rules prevented South Australians from buying water for critical human needs, such as in the event of a supply shortfall.

In June 2011, South Australia's Rann Labor government agreed to settle the High Court case with the Victorian Baillieu Coalition government, which affirmed its removal of the 10 per cent absolute cap on trade out of a water district, and committed to removing the 4 per cent annual cap by 30 June 2014. The parties also agreed to sign immediately the relevant Murray–Darling Basin Agreement schedules that provided South Australia with permanent rights to store water in upstream storages such as the Dartmouth and Hume dams.

These decisions, combined with the water reform packages of 1994 and 2004, meant it was possible from 2014 to trade water with your neighbour, between valleys and between states.

Some restrictions, however, remained, including on inter-valley trade to and from the Goulburn and Ovens rivers, and on inter-region trade through the Barmah Choke.

For the water market and its managers, the Choke presented major challenges. During spring and summer periods of peak water demand, for example, it limited the delivery of traded irrigation water. To manage the flow of water through the Choke, a default trade restriction was established. According to that restriction, a water trade could only take place from above the Choke to below the Choke when there was sufficient offsetting trade in the other direction. (This is referred to as 'back trade'.) This and other restrictions caused differences in the price of water rights above and below the Choke.

The water market design allowed trade between the top and bottom of the system, but the water market designers expected in particular that the water rights would flow predominantly upstream. There, they thought, was where the marginal unit of water would have the greatest value. Upstream, the soils were better, there was more irrigation infrastructure, there was more complementary infrastructure (transport links, towns and regional cities), there was a much higher degree of value-adding of agricultural products, and the water rights upstream were closer to the head of the catchment, and therefore had greater certainty: the water yields were expected to be higher.

In addition to limiting trade geographically, Basin governments had limited the extent to which outsiders—people and corporate interests who did not own land in the Basin and did not farm there—could participate in the water market. In 2004, for example, the Victorian government announced it would introduce a 10 per cent limit on entitlement ownership by outsiders in each Victorian sub-system.

'There is concern in the irrigation community,' the government stated in 2004, 'that non-irrigators could buy up much of the water and drive up its price. The Government believes this risk is more imagined than real. No water will be available to buy unless irrigators choose to sell. In the long-term, the price of water will be based on the value people generate from actually using it.'

The government did not expect the 10 per cent limit to be reached in the near future. 'All the permanent trade that has ever taken place in the 12 years since it began has not yet amounted to 10 per cent of entitlement. Moreover, much of the permanent trade will continue to be from one irrigation business to another.' The non-user limit came into effect on 1 July 2007, when unbundling occurred.

A subsequent Victorian government review in February 2009 found that the limit was an unnecessary—and potentially harmful—constraint on trade. Based on that finding, the government introduced a bill to amend the state's *Water Act 1989* to remove the non-user limit. Water Minister Tim Holding told the Victorian parliament in 2009 that the bill would give irrigators and other water rights owners greater freedom to deal with their rights, and more flexibility to respond to drought:

> The non-water user limit was introduced to allay concerns expressed by some people that, following the 'unbundling' of water entitlements to improve water trading, non-irrigators, or so-called 'water barons', would enter the market and drive up the water price. The government reviewed the non-water user limit [and found] that the non-water user limit, once reached, would constitute a barrier to trade and that fears of 'water barons' significantly entering the market had not eventuated in jurisdictions without a non-water user limit.

Minister Holding further noted that, 'where investors have bought water shares, they have been trading seasonal allocations to users, so the water remains in the market'. The Victorian government's removal of the limit became law on 16 September 2009. This was consistent with the prevailing free-market ideas, as was the 2010 submission from the Bankers' Association to the ACCC:

> ABA agrees that the Basin Plan Water trading rules should provide that there are no specific restrictions on ownership provided the access rights have been unbundled from use...ABA agrees that water access rights should be tradable without the need of the purchaser to hold rights relating to the use of water where there are separate processes for this purpose...ABA agrees that purchasers of water rights should not have to be landowners.

Guided by submissions such as that one, the ACCC gave the green light to external participation, including by those with no interest in farming and irrigation. As the ACCC determined:

> The Basin Plan water trading rules should provide that there are no specific restrictions on the trade of water access rights based on whether any party to the trade is a member of a particular class of entities, including entities such as: non-landholders (unless use has not been unbundled from water access rights); environmental water-holders; urban water authorities; foreign entities (except to the extent of the operation of the *Foreign Acquisitions and Takeovers Act 1975*).

Despite the prevailing momentum towards an unrestricted market, there were further warnings from some quarters about the possibility of negative impacts from an entirely free water market. Professional economists and market design experts shared horror stories of how some of the first experiments with using markets in public policy had run off the rails. In telecommunications spectrum auctions, for example, bidders had colluded to carve up the spectrum between them, rather than pay a competitive price.

(A water exchange manager later rejected the idea that the overall water market had been 'designed' at all. 'Actually there really weren't any "market designers" as such,' he said. 'The market really grew through happenstance. The policy process was driven at the intergovernmental level, without much thought to how the market would or could work in practice. It was more of a guess than a deliberate strategy. No one ever defined what a "successful" water market would look like.')

Other market experiments, too, had failed, and farming groups feared water trading could lead to wholesale corporatisation and industrialisation of Basin agriculture. Some rural and regional politicians echoed those fears, but such misgivings about rural impacts were summarily dismissed. Regulators and senior government officials argued the farmers and the National Party misunderstood how the market was going to work. They reassured the farmers that efficient, well-run farms would thrive, and the market would be a boon for the foodbowl regions on the Murray and Murrumbidgee. Everything would be okay, they promised, when Australia pushed 'Go' on the free market.

Thus validated, the market designers pressed the start button on Australia's gigantic water experiment. Local pundits lauded the Basin governments for leading the world in the development of water markets. The National Water Commission echoed that praise, stating in 2011, 'Australia's experience of introducing a market-based mechanism for allocating scarce water resources could be instructive for other nations dealing with the challenges of water scarcity, as well as in the management of other natural resources.'

Policymakers in those other nations were amazed at how far Australia was willing to go.

CHAPTER 8
DEAD WATER

The seminal work of the modern environmental movement was focused on water, and the health of riverine ecosystems. Rachel Carson's 1962 book, *Silent Spring*, exposed the catastrophic impact of widespread and indiscriminate use of chemical pesticides and insecticides. It was an important catalyst for the establishment in 1970 of the US Environment Protection Agency (EPA) and—in the same year—the Victorian EPA, Australia's first. Governments and community leaders became more and more attuned to pollution and the loss of biodiversity. They were increasingly attuned, too, to how the overuse of rivers and creeks could hurt the environment.

In 1975, the Victorian government initiated two major investigations into the environmental impact of irrigation. The investigations spawned a trio of reports, in 1982, 1984 and 1986, that concluded the water made available by the Dartmouth Dam had largely been committed, 'and therefore only a small volume (approximately 4 per cent) should go to increasing supply and that generally no new entitlements should be given to irrigation'.

In 1989, Victoria introduced a direct water allocation for the environment. This was legislative recognition of the environment's water needs. River managers would allocate environmental flows deliberately, rather than treat such flows as an afterthought or leftover from irrigation and other economic uses. (Farmers called the environmental flows 'duck water' and 'goanna water'.) Allocating water to the environment became a pillar of Basin water policy.

The summer of 1991 was the catalyst for another awakening. A toxic bloom of blue-green algae, a thousand kilometres long—one of the largest algal blooms ever recorded anywhere in the world—clotted the Darling River. The bloom was a portent. During the final decade of the 20th century, the Basin would experience drought, salinity and general land degradation, as well as reduced river flows and poor water quality.

In 1992, the year of the Rio Declaration on Environment and Development, the Murray–Darling Basin Commission was established with a broad mandate over Basin water management. Queensland joined the commission the following year, and in 1995 a historic decision was made to limit overall water diversions. The Murray–Darling Basin Commission placed a temporary cap on extraction of water from the Murray. The cap caused a spike in temporary water trading, and it became permanent in 1997. This was a major turning point, but the Basin was far from rescued.

The year 1998 was the hottest on record (that record has since been broken several times). Climate alarm bells were ringing. The Rio precautionary principle was suddenly urgent: 'In order to protect the environment, the precautionary approach shall be widely applied by States according to their capabilities. Where there are threats of serious or irreversible damage, lack of full scientific certainty shall not be used as a reason for postponing cost-effective measures to prevent environmental degradation.'

Basin rainfall had been low since 1996. But by 2001, the Basin was demonstrably in a severe state of drought—and the drought would last more than a decade. In 2003, the participating governments agreed to find more flows for the Murray by rolling back irrigation and other economic diversions. That agreement resulted in provisions of the 2004 National Water Initiative that elevated the sustainable use of our water resources.

Irrigation lobby groups were sceptical to begin with, but many came round to supporting and in some cases advocating the 'cap and trade' system for Basin water. Environmental groups, too, embraced the system as a way to protect the environment. They also embraced the use of the same approach in other spheres, such as the management of aggregate greenhouse gas emissions. Between left-leaning greens and more conservative browns, there was a remarkable buying in to the cap and trade approach.

The combination of a cap and unbundled water rights did, however, have some perverse effects. With the introduction of water trading, unused entitlements that conferred rights to take water, but which were not being used, or were only being used intermittently—so-called 'sleeper' and 'dozer' licences—were suddenly woken up and activated. This increased water scarcity across the system, as the National Water Commission noted in a 2011 report:

> holders of such licences realised that they had an asset that was of value to those who needed additional water.

> Initially, therefore, water for new developments was largely sourced from unused (sleeper) or underused (dozer) licences, rather than from existing uses, leading to an increase in aggregate water use.

Rising salinity was a consistent concern in the Basin. The Basin Salinity Management Strategy, introduced in 1988, required each state to account for impacts of water trading. In 2002, the Victorian government introduced a salinity zoning policy that defined areas based on the degree of salinity impact from irrigation, and imposed water-use and trade rules to limit adverse impacts. The idea was that farmers wanting to buy water from a lower-impact area would have to pay a levy, and this discouraged water trades that would increase overall salinity.

Also in Victoria, a Water Act amendment of 2005 established the inaugural Environmental Water Reserve, which legally protected water for the environment. Environmental entitlements were now legal rights to access water at certain storage reservoirs and to release it into waterways to mimic natural flows. The amendments included stream-flow management plans and local management rules, which set conditions on when river water could be taken.

For the first time, Victoria carved out a portion of the available water to ensure environmental and ecological benefits. The redgums would get a drink. The fish and ducks and goannas would survive. In 2011, a further amendment established an independent statutory body, the Victorian Environmental Water Holder, to hold the environmental water entitlements on behalf of the environment.

Across the Basin, there had previously been bipartisanship and interstate cooperation in river infrastructure investment. Now, another high point in collaboration was reached. Passed under the Howard government and implemented under the Rudd–Gillard governments, the *Water Act 2007* turned the Murray–Darling Basin Commission into the Murray–Darling Basin Authority (MDBA), and gave a foundation to the Murray–Darling Basin Plan.

Malcolm Turnbull was federal water minister when the legislation was passed. 'Turnbull introduced the 2007 Water Act,' farmer and barrister Robyn Wheeler said, 'on the basis that it was going to save the Murray and on the basis of the Ramsar Convention [protecting the Coorong]—that's how he got it through constitutionally, because he wouldn't have got it through otherwise. Through a section of our Constitution, believe it or not, on "external affairs". It was to get the green vote in Melbourne and Sydney—people who buy, like I do, Earth Choice detergent, because you feel good. As you can tell, probably by now, I'm fairly cynical of politics on both sides.'

In 1983, Bob Hawke had used that same constitutional power to stop the Tasmanian government building a hydroelectric dam on the Franklin River. Hawke had campaigned and prevailed against Malcolm Fraser (who'd also attempted to stop the dam, though by offering compensation) on a promise that he would intervene and prevent construction on wilderness protection grounds. In a landmark judgment, the High Court found the Commonwealth did indeed have suitable powers, in section 51 of the Constitution, to stop the dam.

In December 2007, Kevin Rudd became prime minister and set about implementing the National Water Initiative. Water was so scarce, and flows so diminished, that the Murray mouth closed at the sea. 'Australia's largest river,' Michael Cathcart wrote in 2009, 'had not flowed at its mouth, without the assistance of dredging, for a decade. The waters in the adjoining Coorong wetlands were so diminished, and so toxic with salt and acid, that environmental scientists were declaring that only a mass-transfusion of water would save that idyllic region from annihilation.'

Ken Matthews, chair and CEO of the National Water Commission, wrote in a 2007 assessment of the Initiative: 'Overallocation of water resources continues to be a central national challenge. It is still not being managed as envisaged under the NWI. A number of states have not delivered on their commitment to move to sustainable levels of water extraction.'

The aim of the Murray–Darling Basin Plan was to improve the health and sustainability of the Basin's river systems, while continuing to support farming and other industries. The inaugural chair of the MDBA, Mike Taylor, later resigned, reportedly because he baulked at putting environmental considerations ahead of economic and social ones when managing Basin water resources.

In 2008, the Commonwealth government announced it would enter the water market, on behalf of the environment, and spend over $3 billion to buy water and water licences. 'Buybacks' would become a dirty word throughout the Basin.

National Party senator Barnaby Joyce told parliament:

> We have had some $23.75 million of this nation's resources put towards the purchase of Toorale Station so that about 14 gigalitres of water can be removed from the station at this time. It will probably go merrily down the river for about 100 kilometres or 200 kilometres—we do not know; we do not even know if it goes into a charged system—and will quickly dissipate. There is an acknowledgment that with this purchase that it is probably not going to do very much if anything at all to

> relieve the pressure on the Lower Lakes and the pressure on Adelaide's water requirements.

The Murray–Darling Basin Plan was signed into law, with cross-party support, in 2012. The Commonwealth, the four Basin states and the ACT collectively established an environmental water recovery target to ensure there was enough water for native trees, native fish and other key assets of the natural environment. The crux of the Plan, implemented by water minister Tony Burke under the Gillard government, was to recover 2,750 gigalitres to prevent the type of ecological collapse that nearly occurred during the Millennium Drought. Implementing the plan, and keeping all the parties at the table, came with a $13 billion price tag.

Subject to review and revision through a seven-year implementation phase, the Plan established sustainable diversion limits and a framework for water resource plans (to be developed by the Basin governments), which set the rules on how much water could be taken. New South Wales agreed to share water with downstream states when volumes in the Menindee Lakes reached certain levels.

But there was a fundamental problem with the Plan. At its most basic level, it didn't take climate change into account. As former MP Tony Windsor later explained, in the aftermath of Tony Abbott's 'carbon tax' attacks, the Gillard government 'almost became paranoid about the word "climate". So [the Basin Plan] had an inbuilt frailty in it.' This became evident as the environmental measures in the Plan failed to protect the state of the Basin.

The Plan's signatories twisted and turned in a series of attempts to fix the flawed roadmap. In 2013, the Commonwealth assigned to the environment a further 450 gigalitres of water, purportedly linked to savings from on-farm water efficiency projects. The following year, the Basin environmental watering strategy was agreed. It advised Basin stakeholders on how to plan and manage environmental flows. Further environmental watering plans would follow, along with revised recovery targets, diversion limits and advice on what a healthy river system needed. In 2015, Commonwealth surface water purchases were capped at 1,500 gigalitres and more latitude was given for water efficiency measures.

But despite all the official twisting and turning, the Basin's natural waterways continued to suffer.

The *Australia State of the Environment 2016* report assessed the Basin as 'very poor' and deteriorating for 'inland water ecological processes and key species populations'. The Wentworth Group of Concerned Scientists, an independent group working to secure the health of Australia's land, water and biodiversity, concurred. There was no evidence of long-term Basin improvements, or that the long-term deterioration in key indicators of river condition had stopped.

In June 2017, researchers from the University of New South Wales found a dramatic decline in Basin waterbird populations; their numbers were down around 70 per cent over the preceding three decades. The Basin started to dry up again and the politics got very heated. Labor and the Greens used the balance of power in the senate to block changes to the Murray–Darling Basin Plan that would cut the amount of water returned to the environment in the northern basin by 70 gigalitres.

In the summer of 2018–19, the rivers, lakes and billabongs of the Basin were parched, oxygen levels dropped, temperatures soared and Australia witnessed piscine megadeath in the Lower Darling at Menindee. In the first die-off, on 15 December 2018, tens of thousands of dead fish were reported. A second fish-kill three weeks later left hundreds of thousands of dead fish in the same stretch of the river.

A viral Facebook video from 8 January 2019 featured two blokes knee-deep in green muck and holding dead cod. 'This has nothing to do with drought,' grazier Rob McBride said. 'This is a man-made disaster.' 'You have to be disgusted with yourselves,' Dick Arnold said. 'You politicians and cotton growers.'

A third event, on 28 January, killed millions more fish. The noble Murray cod is an iconic species: if the Australian coat of arms featured a fish, it would be one of these. But the gasping and rotting fauna at Menindee included hundreds of Murray cod, some over 50 years old and weighing 40 kilograms.

Tempers again flared and fingers were again pointed. An investigation by the Australian Academy of Science into the causes of the mass deaths concluded: 'The conditions leading to this event are an interaction between a severe (but not unprecedented) drought and, more significantly, excess upstream diversion of water for irrigation. Prior releases of water from Menindee Lakes contributed to lack of local reserves.'

Opposition spokesperson Tony Burke told the ABC's Anna Henderson that the allocation of an additional 450 gigalitres of environmental flows to the river system 'was bipartisan until Barnaby Joyce threw it all under a bus'.

'It took us one hundred years to reach this stage,' National Party leader Michael McCormack said of the Plan on ABC radio. 'Of course there needs to be some tweaks.'

In what seemed to be a further fracturing of bipartisanship on the Plan, Tony Burke announced Labor wanted to remove the 1,500-gigalitre cap that was established in 2015. The cap restricted the amount of water that could be bought by the Commonwealth from willing sellers to return more water to the environment. National Irrigators' Council CEO Steve Whan said, 'It seems to be a breaking down of the bipartisanship, which is vital to the success of the Plan.'

Scientists working in the Murray–Darling Basin resorted to medical metaphors to describe what was happening. Wetlands scientist Richard Kingsford likened the Macquarie Marshes to 'a very ill patient who has been given just enough care to get out of the ICU'.

People living on the Darling went even further. Barkindji Elder Lilliana Bennett recalled her grandmother talking about going down the Barka to fish and hunt for goannas.

'It's a place they go to relax, to tell stories,' she told *SBS News* in 2019. 'For me, it's been really devastating, I mean, we went down and camped by the river where there's still a bit of water around and it just doesn't have the same feeling. It's dead water.'

CHAPTER 9
A WINDFALL

'I grew up on my father's irrigation property,' farmer Laurie Beer said, 'which is just down the road from where I am now. I went to year 10 and then I left school and worked for him for about 12 months. And then I share-farmed with him for several years until I got married and then I bought my own place, which is mixed irrigation.

'Rice was always our key crop. The other things were just sort of a sideline but it's turned around now and rice has become an opportunity crop because of water availability and the price of it. I've been concentrating on sheep the last few years and I grow rice when water's available. I grow a bit of winter cereals as well, a bit of hay sometimes, just whatever I can.

'We raised four kids here on our home property since we've been married. My son now works with me on a share basis. And the three daughters have all got jobs independently. I haven't done a lot of buying of water in the past. I try to run my operation, as I call it, as a bit of a shoestring operation. I try to keep the costs to a minimum.

'The property I bought here basically had two standard water rights with it—what was considered standard at that time, which meant I had a pretty good allocation back when I bought it. I really didn't think I'd ever buy water temporarily, because I had the allocation there—that would have been enough to do what I wanted to do—up until the Murray–Darling Basin Plan changed things.

'I had a surveyor in to do a farm plan not long after we bought the property and he asked me how much water I had and I told him, and he sort of looked at me and said "Oh, you'll never use all that." In those early days it was actually a couple of times where I had a bit of surplus and I did sell a little bit, just as extra pocket money because I really didn't need to use it.

'So I've been a bit slow to get into the thought process that I have to buy water to carry on our operation. But I've come to the realisation that the only way I'm going to remain productive, to do the things I want to do, is to become a bit more aggressive and buy water when I think it's cheap enough.'

By 2000, Australian water trading was an oldish concept, but fewer than 10 per cent of Basin irrigators had conducted a water trade. From 2004 onwards, however, temporary water trade accelerated as the National Water Initiative reforms were implemented. By 2015, the proportion of southern Basin irrigators who had conducted at least one water allocation trade had grown to 78 per cent. Permanent trade increased, too, particularly after the Millennium Drought.

Enthusiastic reformers at the PC and the ACCC—and government departments such as the Commonwealth departments of treasury and agriculture—continued to encourage farmers to embrace water trading. 'The reformers,' Michael Cathcart wrote, 'talked of irrigators trading water up and down the river as if they were so many shopkeepers swapping goods along the high street.'

Banks, too, encouraged farmers to sell their water. Reporter Kath Sullivan said on the ABC *Signal* podcast in 2019: 'A lot of the people who have irrigation agriculture along the Murray–Darling Basin system have actually told me stories about how their banks advise them to, you know, just get in there, get going and get trading on the temporary market and worry about getting a permanent allocation a little bit down the track…The advice at the time was to, you know, "get your capital into your crop".'

Tradable water rights became a significant financial asset for many farmers. The water could be sold virtually anywhere in the southern connected Murray–Darling Basin. There were willing buyers and sellers at all times.

In 2007, a report into the impacts of water trading was jointly commissioned by three government bodies: the Rural Industries Research and Development Corporation (later known as AgriFutures), the National Water Commission and the Murray–Darling Basin Commission. The experiences of many farmers, irrigators and community members were captured in this AgriFutures report.

'In our district,' a farmer near the Victorian town of Boort said, 'we have two very distinct soil types. One is just hopeless without irrigation water; that's our heavy black clay. Where a lot of water has been sold from is the red soils; they can farm them dry. But on the black soil it takes so much water just to get them wet.'

'Water trade has provided many farmers with a form of superannuation,' a farm manager said. 'Yeah, well it allows them to live… Their investment in land and water is more liquid in that they can sell off part. Generally farmland is not very liquid: you have to sell the whole business or sell nothing, whereas you can sell off a portion of your water relatively easily at the moment.'

The unbundling of water rights created a windfall for some farmers—such as those who owned marginal land around Echuca and Nathalia in northern Victoria. Suddenly these farmers had a significant, tradable asset, and there were disputes within families over valuable water entitlements. In many cases, the water was more valuable than the land.

Economist Michael Woolston wrote in 2005:

> [T]he unbundling of water from land titles, while a critical step to allow water trading, gave rise to a number of financial, legal and related issues that had not been fully anticipated under the COAG reforms. In particular, while the combined value of land and water should in principle increase when the water is able to be traded separately, much of the value may rest with the water rather than the land. In fact, the value of the land, without the water, may fall significantly.

The impacts of differential land and water values played out in several ways: via loans secured through mortgages over land; through bequests of land and water entitlements in wills; and in the diminished rates bases of local councils.

The situation for many Basin farmers considering whether to sell their water was variously described as a game of chicken, a Mexican stand-off and a prisoner's dilemma. There was a gain, to be sure, but it soon dissipated, and farmers could only sell their permanent water once.

'When I first became a water broker,' former real estate agent Desiree Button said, 'it was just a natural fit, when they separated land and water. When I first started off in the job in about 1996, obviously land and water were together, and then in 2007 they separated them. That was wonderful at the time because everyone thought their asset base would increase considerably—which it has. There were lots of positives, and then the horse bolted and you can never turn back.'

Water trading, many farmers hoped, would bring extra flexibility to how and when they used water. It meant they could sell water they didn't need; they could augment supply, especially in times of drought; and they could turn their water rights into a monetary asset. Those who sold all their permanent water had to re-enter the market each year to buy temporary water, provided they wanted to continue to farm.

For farmers like Jerilderie's Peter Burke, water markets were an opportunity to manage risks, grow the business and keep food on the table.

'The only reason we got involved in the water market was purely for farming reasons,' Peter said. 'My wife and I bought our first two farms when I was 22, in late 1999, and we had about 1,400 megalitres of permanent water. I'd grown up on my parents' farms. They are down the road and we always saw them, planning around every season, different allocations. Some years you have drought, no water. Either feast or famine.

'So we started off our farming business the same way. Renee and I bought our first two farms on vendor terms, off a fellow who I was working for. So we were in debt up to our eyeballs and every year would come around and you would end up with no water and no income. So right from the start, I always said to myself, "How the hell can we stop the peaks and troughs? It's just ridiculous." We had about a million dollars tied up [in] water. Couldn't find or buy more land.

'All the equity was tied up in the permanent water, and the yield was so unreliable—a bit some years, none others, all of it the year after that. Then came the Millennium Drought. We'd bought machinery and we were share-farming and all sorts of stuff. And then we were basically left holding the baby when the drought hit. We had all this debt and no income. Along come the government making offers to buy permanent water, to take it back for the environmental flows. And, gee, it was nearly double, might have been more than double, actually. About $1,400 a megalitre.

'Back in 2003 or '04, I was only making about $680 per meg, and we had a million-dollar debt. I was buying four and five thousand megs of temporary water every year to grow thousands of acres of rice and winter crops and whatnot. And then the banks said we'd either have to sell everything or sell the water. I kept saying to Renee...because it was sort of a bit of a taboo subject, you didn't separate the land from the water. It was frowned upon back then, whereas now it's done every second day.

'So we were forced into thinking, "Right oh, well, I don't want to change the way I'm farming." So we cleared our debt by selling the water, and then we went and bought more land, which cost us nothing, and then sold the water from that land as well. That was very handy.

'It's too risky to go and grow thousands of acres of rice when you can just, you know, pocket the money and pay off the debt. But we were able to keep growing small areas of rice. So income-wise, instead of going into a drought and thinking, "Oh no, we're going to go broke," because of what we did, actually we got further ahead quicker than what we were.'

Laurie Beer said: 'People are pretty reluctant to talk about their operations as far as water goes. I don't quite understand some of the logic behind why people are so secretive about what they are doing. Some people think it's a sin to sell any water at all. I don't hold that view. I hold the view that if you need to do it, well you do it. But you don't hide it. You've got a budget to meet, and if you need to, you sell a bit of water to make ends meet. Selling it might be a better option because you haven't got enough to grow something with, so you do it. Well, that's fair enough. I don't see a problem with that. The same thing if you need to just buy a little bit more, well that's what you do.'

The 2007 AgriFutures report revealed some sobering stories about the impact of water trading. Some farmers who chose to sell permanent water were criticised and even ostracised in their communities. The sale of entitlements had unexpected impacts on irrigation schemes and rural towns, as one farmer explained.

'There is a chap here who sold all his water,' said the farmer. 'I don't think he's even got stock and domestic. That 320 acres goes out of production...I don't think he's going to be able to live there so he'll move out and that's a loss of kids from the school, spending in the town—that's all because of water trade. Ten years ago he would have sold the farm and the water would have stayed with the farm, and someone else would have bought the farm and had a go.'

A business owner in Pyramid Hill noticed a lot of dairy farms were shutting down. 'Years ago, I probably traded 50–50 with dairy farming and dryland beef and local businesses, and it would lucky to be 15 per cent of dairy now, largely due to deregulation and farms downsizing. But the biggest influence is water. People have just sold water and got off the farm. Most of the dairy farmers that have sold have either stayed on the farm and are dryland farming or they are selling and moving to other areas, and along with that goes their share farmer and workers. So far, business is stable. After Christmas it might change. If the conditions are the same we will be reducing staff. It's hard to keep positive.'

The impacts of permanent water sales had significant flow-on effects. In the Goulburn Valley, for example, there was 'a general sense of grieving' over the imminent collapse of highly integrated dairy farms and milk factories.

'The whole thing was originally set up as an integrated system,' a farmer said. 'Now a massive amount of water from the Goulburn system has moved downstream of Echuca, and the people in Tatura Milk Products are beside themselves about where the milk is going to come from. And it's only five years ago that dairy farmers would do all sorts of things to become suppliers to that company.

'A friend was showing me in that area a five-mile strip of road where there were only two functioning irrigation farms left. And it is such beautiful country. So much effort went into making sure that the whole integrated system centred on the best soils close to where the water was harvested. Now the water is going hundreds of kilometres away from the storages to areas of poor soils, and then the produce is being transported miles back to the processing works.'

Some farms were uneconomic without permanent water. A lot of farmers decided to sell their rights, take the money and walk away altogether. (Astute players bought cheap land, stripped and sold the water rights, then left the land vacant and unirrigated. In America, that strategy is referred to as 'buy and dry'.) Farmers in financial distress or with significant debt had to sell their permanent water, and this sometimes put them in a downward spiral.

'I've got a mate of mine who I really feel for,' an irrigator said. '...He's had about 250 tonnes of wine grapes, but last year he didn't get one berry in, not one berry. I said, "What's happened, get any for next year?" "Nah, I've got nothing", he reckons. He has this water, which he still has to water his vines, so he is going to cut back on the water to see if he can get some money in with his water trading, which at the moment it's not real big money but he's trying to get some money out of that to keep it going. The thing is, he is just going to go down. I can see him going down.'

Financial distress, combined with the sale of water and the abandoning of farms, had a tangible impact on the landscape. Dormant and deserted farms became havens for weeds and pests.

'At the moment,' another irrigator said, 'people have sold their water off, so they are not irrigating. They are leaving the snails and the birds and rabbits and weeds; it's a different scenario than we are used to.'

'The biggest problem with the place that I'm trying to buy,' a farmer said, 'is that [the seller's] gone out of dairying. He's sold off the bigger percentage of his water, so he's probably got his money, but bloody noxious weeds and thistles...they've become a bloody nuisance. I look at all the dry blocks around me—they don't worry about cutting their thistles any more—and, as an adjoining neighbour, it's a bloody disaster.'

'Because of permanent trade,' an irrigator said, 'people walked off and left the district. The population decreased, income went down with water taken off the land. It flattens everyone. You drive past a beautiful, thriving farm and see it just let go. It upsets people.'

Many farmers had second thoughts about selling permanent water.

'Yes we have sold off,' one farmer said. 'We had 660 megalitres in total and we have sold 100 permanently. I wish I hadn't done that now; it's not worrying me that much but I wish I hadn't sold that much.'

CHAPTER 10
AT A LOSS

Water trading might have been a devil's dilemma for Basin farmers, but many of them embraced it, just as governments and banks had encouraged them to. The reality of the market, however, was a long way from a high-street shopping trip.

Buying and selling water is complex, much more so than buying consumer products or many other types of tradable commodity. Gold, for example, is the same wherever it comes from, and it is extremely easily to sell, whereas water rights are multi-dimensional—as is the whole water market. The Basin water market straddles different valleys, zones, regions and water trading jurisdictions, each with its own rules. There are important differences in 'carryover' rules, for instance, between Victoria and New South Wales.

In day-to-day trading of water rights, farmers' bids and offers vary according to price, quantity, place, type of right, type of water (such as surface water vs bore water) and the likelihood of receiving the water (aka the 'yield'). Different types of water rights are associated with different legal rights. Posted prices, too, have hidden differences and hidden complexity. Additional imposts may include exchange costs and trade approval fees. Farmers in irrigation schemes and irrigation groups also face specific local costs and rules.

Ben Dal Broi grew up on the family farm in the Murrumbidgee Irrigation Area, to the west of Griffith. After completing his rural science degree at the University of New England, he worked for the federal Agriculture Department in a wide variety of policy areas at the executive level, including a secondment to the Department of Prime Minister and Cabinet during the Abbott government's development of the Northern Australia and agricultural white papers.

Five years ago, Ben took his long service leave and moved with his wife to Griffith to help his parents on the farm while his dad recovered from a hip replacement. The intention was to return to Canberra, but something 'clicked' and Ben suddenly realised farming was what he wanted to do. 'The only person who wasn't really surprised was Mum. "Farm boys are like salmon," she said. "They come home to breed."'

Ben told us, 'When water is bought or sold, there is a plethora of fees tacked on to the transaction. There are the broker's fees of between 2.5 to 4 per cent, which, depending on the broker, falls on the seller or is split between the buyer and seller. There can be a fee from the irrigation companies managing the network you are on. There are any number of state government charges such as property securities register searches, land title searches, administration and compliance fees, and additional fees if you are transferring water within, into or out of valleys or irrigation districts. They mightn't look like much, with a couple of hundred bucks here and there, but before you know it you can be paying thousands a year in fees.'

As a consequence of the constitutional power of the states over water, each Basin state has its own water legislation, and these create the differences in rights, along with differences in how those rights are traded, and how trades are settled and approved. Across the Basin states, as well as having different legal fine print, water rights have different names. Victorians, for example, speak of 'access entitlements', 'water shares' and 'take and use' licences. In New South Wales, you will hear 'water access licence'; in South Australia, 'water access entitlement'; and in Queensland, 'water allocation'.

State and federal government bodies, plans and agreements add to the complexity. There is a gaggle of public entities and regulators—including the state water and primary industries departments—and a whole bookshelf of policies and plans, each with its own complexities and controversies.

'The Basin Plan is a contentious issue,' farmer Chris Brooks said. 'There's a lot of ambiguity about rules, volumes, what has happened over a long time with successive state and federal governments. We live it every day and we've been saying the same thing. It really is a mess.'

'We own the water,' a Basin business owner said in 2007, 'not the government, but they keep changing the rules and the structures.'

All farmers must navigate through the technicalities of yields, allocations, carryover, grandfathering, diversion limits, 'over-bank transfers', and rules for inter-valley transfers, such as between the Murray and Murrumbidgee. Some irrigation, such as unregulated 'floodplain harvesting' in the north of the Basin, even falls outside the 'cap and trade' rules altogether. ('Floodplain harvesting' refers to the use of earthen banks and levees to collect floodwaters into large private dams and storages.)

In addition to buying and selling water on the 'spot' markets, for near-term delivery, there are 'forward' markets in water, for future delivery, as well as embryonic side markets, where people trade water derivatives such as water options, which are a kind of wager on future water prices.

Farmers must also decide whether to buy and sell 'on-market' or 'off-market'; and if on-market, which trading platform and water exchange to use. To secure enough water, farmers often have to buy it across multiple exchanges and platforms. There are public and private exchanges, along with public and private water information services and matching apps such as Waterflow and Waterfind. At a recent New South Wales inquiry, a parliamentary committee member called one of these apps 'the Tinder of the water industry'.

Farmers must decide whether to commission specialist advice from a water planner or consultant, and whether to seek the services of a water broker. In practice, a large proportion of water trades take place via brokers, and many of those trades are executed offline and outside the water exchanges. Some trading also occurs through direct negotiation between farmers, just like it did in the old days.

'I have a private franchise arrangement,' water broker Desiree Button said, 'as a real estate and water broker in Echuca and the wider area. I sell and trade both temporary and permanent, surface and deep well water. Have been doing that for about 25 years. Long enough to know better! Water is a lot more complicated and nuanced than selling land. Just think of the vast geographic areas, the different types of water, all the different rules, the multiple trading platforms and the maze of bureaucratic bumf.'

'Water literacy has changed,' interim Basin inspector-general Mick Keelty wrote in 2019, 'such as the need for irrigators to now understand and have knowledge of how a water market operates.

A farmer needs to not only farm, but simultaneously be an economist, trader, and weatherperson. There is limited time to be able to do all of these things, resulting in some family farms falling behind.'

'In 2006,' farmer Shelley Scoullar said, 'wheat had become deregulated. Before then, you just sold your wheat to the wheat marketing board and that was that. You didn't have to do any trading. Then all of a sudden, farmers had to market their grain, and do all the research on the best way to do that. Do they use forward contracts? Do they warehouse it and sell it later? So many decisions. Then, all of a sudden, they've got to be trading water as well.

'Most farmers just want to be out in the paddock and making the best choices and the most efficient decisions to grow good-quality, high-yielding crops or cattle, whatever it is. And then all of a sudden they've got to be indoors making these trading decisions and marketing decisions.

'Some, if they had the capacity, might have gone off and learnt about water markets and what sort of products to use. But if they didn't have that capacity, they would still do what they had always done, which is probably me and Dad, because water markets were too much for me. Dad actually had depression at the time and I was handling all the grain marketing for both families. That was enough pressure for me. And plus three kids. I had plenty going on! So I can see—and it's easy to see if you are a family farmer—how that whole water marketing space is just too much. There were so many different products all of a sudden.'

The most important difference in water rights is the difference between temporary and permanent water. Here, too, there is important fine print. A 'permanent' entitlement does not necessarily mean guaranteed or even reliable water.

'Just because you have a permanent bucket, doesn't mean that you're going to get any water,' reporter Kath Sullivan said on the *Signal* in October 2019. 'This year in New South Wales, southern New South Wales irrigators have got their second year of zero per cent allocation. That means irrespective of the bucket that you own, the "bucket asset", if you will, they can't fill it.'

Before the Millennium Drought and the water reforms, Ben Dal Broi's family reliably received 100 per cent of its 'general security' allocations—a predictable supply of water with which to operate a modest farm business and build a future for his family. At the turn of the century, though, things began to change. Since 1999, the farm experienced a sustained decline in water allocations by at least 40 per cent, and often by 90 to 95 per cent. Ben now refers, ruefully, to 'general security' water as 'no security'. Like most of his neighbours, he has taken on debt to buy additional entitlement and temporary water—not just to replace the shortfall but to survive.

Over recent years, the market price of permanent water has gone from around $1,000 or $2,000 a megalitre to more than $8,000. As Waterfind's Tom Rooney explained in an interview for ABC radio, 'We're seeing now high-security water prices at $8,700-plus per megalitre, and those same water entitlements in 2007 might have been worth $2,000 to $2,200.'

Temporary water, too, is more expensive, though there is more variability in temporary water prices across seasons and years, and even simultaneously across valleys and districts. Rooney gave an example of how, in November 2019, temporary water prices differed across some key water markets such as the Goulburn, the Murrumbidgee, the Victorian Murray and the South Australian system.

'So if we go through some of these districts,' he said, 'such as Goulburn 1 sitting at $650 currently, versus zone 6 on the Murray in Victoria sitting at $640, versus zone 7 in Victoria sitting at $900, and then we go to zone 10 General Security, which is above the Choke at $640 [and] $900 below the Choke in South Australia. And New South Wales, where we've got then a Murrumbidgee price sitting at $745, and there's a variety of different prices there.'

Even the most well-informed farmers struggled to fully understand the water market, or to optimise their returns from buying and selling water.

'You have to be pretty smart to pick the peak in the market,' one farmer said in 2007. 'It's not an easy market to read because they have put so many rules in place, and the restrictions on where you can trade. The information is there, but you've really got to… Well…My gut feel is buy locally, sell out of town. That's what I've been doing.'

Peter and Renee Burke took a cautious approach to buying and selling water.

'I don't muck around with it too much,' Peter said. 'In a year where we are selling, Renee and I sit down. We'll go, "Right, we paid X for it. It's now worth this much. When it hits whatever trigger point, we just get rid of it." We don't take the risk because it can rain tomorrow and the dams are full and all of a sudden water drops, you know, half price. So I don't try and hang out for an extra massive amount of dollars, not big money anyway. You don't really take that much risk with it. We just pick a direction and run with it.'

One option for farmers looking to 'cut through the complexity' (to quote a major accounting firm) was to trade via water brokers. According to the original water market design, different types of brokers would stand between buyers and sellers of water rights, to help them connect and transact, either on exchanges or 'off-market'. The role envisioned for brokers was a bit like how a real estate agent might help people find and buy or sell property. And that analogy is especially pertinent, as a lot of broadacre real estate agents, including Desiree Button, became water brokers when trading started. By 2020, the population of Basin water brokers numbered in the hundreds.

Desiree explained the process of water broking to us. 'People ring up, and they might say, "I want X amount of water, how much can you get for me?" Then you probably gauge it against a couple of different exchanges, like MIL [Murray Irrigation Limited] for example, and RuralCo and Waterfind for that matter. There are always benchmarks. It's fairly easy to follow. I do the deal. I'm the middle transactionary person.

'So if someone rings up and says, "I want 100 megs. How much is it worth?" Depending on where it was, I'd say, "It's worth 100 bucks per meg on MIL." If it's below the Choke, add $100 or $80 depending on what the market is doing on any particular day. The market can change $10 or $20 depending on a rainfall event, or people punting on what DPI [the New South Wales Department of Primary Industries] may or may not allocate the following day, on the first and 15th of every month. There are just a lot of variables. It can be a week of hot weather coming that usually pushes the price up.

'Once I've set the price, I lodge the transfer forms, send a tax invoice, get paid, and once it is paid I submit the WaterNSW trade forms. As soon as it is settled, I pay out the vendor.'

Things were somewhat different for water broker Shane Sutherland. 'Most of my transactions are off-market,' he said. 'I don't use the water exchanges much at all. I do use the RuralCo one sometimes. But, I'd say 95 per cent of my trades are offline and certainly all my permanent sales are offline.'

Faced with a highly dispersed and extremely complex market, Basin farmers were often at a loss. Many of them had minimal IT expertise. Typically they relied on off-the-shelf and obsolete hardware, such as cheap and fully depreciated laptops from Harvey Norman and Officeworks. Many farmers did not have reliable broadband—or any broadband. For time-sensitive water trades, they relied on 1980s and 1990s technologies such as landline phones and email.

'I remember,' Shelley Scoullar said, 'during the 2014 rice harvest, by then, Dad and I had realised we needed to get more active in the water market. We hadn't really purchased much water off the exchange, but we realised to ensure we could grow a rice crop the following year we should try to purchase temporary water to carry over into the following year. Bearing in mind we still didn't have a lot of capital on hand as, coming out of the Millennium Drought, we were keen to put a dent in some debt.

'There must have been an allocation announcement at around the time, as we were watching the temporary water prices. I was on the chaser bin and in between loads I would stand on the top of the tractor roof to get better reception on my phone to check the prices. Then I would get on the UHF to discuss with Dad and decide what to do. For a young couple who had two small kids, just coming out of the Millennium Drought, without much cash behind them, even spending $5,000 on water was a big decision. Then we had to race home to use the dial-up internet to print the forms to buy the water. It was stressful.'

Laurie Beer traded water through an old laptop on his kitchen table, connected to the market via a WiFi connection that kept dropping out.

'When the hell are we going to fix the NBN!' he regularly exclaimed.

When the water market was first implemented, little consideration was paid to training farmers or equipping them with the tools they needed.

'Many older farmers struggle even to use a smartphone,' Peter Burke said. 'They simply can't use the water trading platforms.' But other market participants certainly could use those platforms—and they would use this fact to their huge advantage.

PART III: PLAYING THE MARKET

CHAPTER 11
BIG WATER

At the start of this century, the idea of 'investing in water' became a full-blown Wall Street fad. (If you are discovering water as an investment now, one insider told us flatly, you are too late.) As Lynne Kirby noted in relation to the 2017 documentary *Water and Power*, the idea of water being a valuable and tradable commodity was first defined and articulated on Wall Street. It was a Goldman Sachs executive who coined the line, 'Water is the oil of the 21st century.'

The 2015 film adaptation of Michael Lewis's bestselling *The Big Short* ends with a tantalising and inspiring line. Investment genius Michael Burry had predicted the 2007 US housing market collapse and the ensuing global financial crisis, but he had since shifted his entire investing focus to a single commodity: water. Michael Cathcart's book *The Water Dreamers* also ends with water trading. 'As the global business magazine *Fortune* told its readers in May

2000,' Cathcart wrote, 'water has become "one of the world's great business opportunities".'

In December 2012, the Montreal-based Centre for Research on Globalization released a list of 'water barons' and discussed the accelerating trend of 'megabanks' buying up water. According to the centre, Wall Street banks and 'elitist multibillionaires' were buying water 'all over the world at [an] unprecedented pace'. A list of water-centred hedge funds from that same year included the Master Water Equity Fund, Water Partners Fund, The Water Fund, The Reservoir Fund, The Oasis Fund, Signina Water Fund (from Switzerland) and the Australia-focused MFS Water Fund of Funds (MFS Aqua AM).

'Being a water broker is a lot like buying and selling shares or any other financial asset,' broker Desiree Button told us. 'Water's a tradable commodity. You've got Pitt Street farmers and non-agricultural representatives from all around the world investing in water. Why wouldn't you? It's a necessary essence of life. And it has a very, very good return on investment. And excuse the pun, but it is very liquid, in terms of the ability to get in and out.'

With water, an investor told researchers in 2020, 'There is no depreciation, there's no goodwill, there is no maintenance and repairs. There is really very few costs in the ownership of it. There are not many asset classes that are that good.'

The promise of high returns and low entry costs reportedly led to a liquid gold rush. Banks, hedge funds, private equity funds and sundry other investors were said to have plunged into water with gusto.

'They raced to it,' a water broker told us, 'like 17-year-olds at the Cootamundra B&S Ball.'

—

We spoke to two water exchange managers on condition of anonymity, and we refer to them here as 'Mr X' and 'Mr Y'.

Are banks and hedge funds, we asked, really trading Australian water?

'Yes they are,' Mr X said, 'but they didn't force their way in. They were invited.'

A central feature of the original water market design, Mr Y explained, was the decision to allow foreign entities and non-water users to participate in the market. Among those external participants, one group in particular would perform a crucial role. External 'investors' would be relied on to provide 'liquidity' in the market.

The idea was that banks and quasi-banks (such as hedge funds and similar institutions) would increase the number of water market participants, and these cashed-up entities in particular would help 'make the market': they would be prepared, at any given time of day, all through the year, to stand in the market and be ready buyers and sellers of water rights.

At the time of the initial water market reforms, Australian banks had been enthusiastic advocates for the laissez faire approach—Mr X called the banks 'cheerleaders', pronounced as two distinct and accented words. During public inquiries and in private meetings with politicians and officials, the bankers had served up conceptual and pseudo-empirical justifications for free trade between valleys and regions, and for the removal of restrictions on who could trade water.

With more market participants, the argument went, farmers would find it easier to buy or sell water when they needed to. This would be crucial insurance for their farming plans; it would help maintain the value of their water assets; and it would lead to a transparent market price that reflected the real value of water—especially if the liquidity providers were market-ready veterans of New York, Hong Kong, Shanghai, Frankfurt and London.

The institutions that entered the market included international investment banks, large investment funds, superannuation funds, and so-called 'family offices', which functioned like small-scale, private investment banks and fund managers, working on an exclusive basis for rich clients. Given an emphatic green light, these institutions entered the water market, just as they entered markets around the world for energy, agricultural commodities, mineral products, stocks, bonds, exotic derivatives and cryptocurrencies.

From the water market managers' perspective, the implications of the decision to allow bankers into water trading were plain to see. In the old days, before 1994, water trading had been a decidedly crusty and agricultural affair. Like many water people, the managers were trained in that era, and they'd entered the water sector with the intention of managing storages and flows, and systems for metering and irrigation.

Deliberately or otherwise, the water market designers had let the banks in as liquidity-providing middlemen. In so doing, they'd turned Australian water into a modern financial product, and the water allocation process into a modern financial market.

This transformation occurred at an early stage in market-oriented public policy. Despite not having the full set of market design tools (including legal and regulatory ones) to guarantee that such a huge experiment would work, Australia's politicians and regulators had turned water into a valuable, widely tradable commodity. In the ear-bothering terminology of Wall Street, water had been 'financialised'. To the hydrology-trained market managers, there was something tragic or even sacrilegious about that fact. And it reminded them, they said, of another troubling experiment.

—

Between the 1980s and the 2000s, the phenomena of financial innovation and FinTech spread well beyond the traditional boundaries of financial markets. There was a massive conceptual widening of what could be 'financial'. Agricultural commodities, for example, became the focus of high-octane trading exchanges, as did formerly dull utility services such as electricity and gas.

Houston-based Enron was the most spectacular early experiment in financialisation. Enron started as an energy company, but it rapidly morphed into a sophisticated derivatives business. When Enron began, its activities were largely unregulated—naturally, because the business was doing something new. It was able to quickly establish a prime position in the new world of tradable energy.

Mentored by Wall Street veterans, including dealers from the investment bank Salomon Brothers, Enron transferred the aggressive tactics of Wall Street trading rooms into energy trading rooms. In the early 2000s, during Californian wildfires, traders on Enron's west coast desk were recorded cheering on the fires, and celebrating the impact on spiking energy prices. 'Burn, baby, burn,' a trader said. 'That's a beautiful thing.'

> 'He just fucks California,' says one Enron employee.
> 'He steals money from California to the tune of about a million.'
> 'Will you rephrase that?' asks a second employee.
> 'Okay, he, um, he arbitrages the California market to the tune of a million bucks or two a day.'

The company benefited from the trends in public policy towards free markets, deregulation and especially privatisation. It controlled energy assets all over the world, including a quarter of the US natural gas industry. At the forefront of developments in market design and FinTech, the company invested in advanced online trading platforms. These became the market standard, and helped turn Enron into the world's main buyer and seller of energy-related securities. In less than a decade, the company established a near-monopoly position in tradable energy.

For six consecutive years from 1996 to 2001, *Fortune* magazine named Enron America's most innovative company. And then it all collapsed into a steaming mess. On 2 December 2001, just two weeks after US Federal Reserve chair Alan Greenspan collected his Enron Award for Distinguished Public Service, Enron's board decided they had no option but to seek Chapter 11 bankruptcy protection for the company that had become one of America's largest financial businesses, employing around 21,000 people.

The end of Enron was also the end of the company's auditor, Arthur Andersen, which was convicted in June 2002 of criminal complicity and felonious obstruction. Stripped of its licences to practise in the US, the Big Five accounting firm was dissolved. A subsequent Supreme Court judgment overturned the conviction, but that decision came too late for the auditor's tens of thousands of employees.

In 2006, Enron's former chairman Ken Lay—a personal friend of President George W. Bush—was found guilty on charges including conspiracy, securities fraud, bank fraud and making false statements. While holidaying in Aspen, Colorado, he died before his sentencing.

Having entered the sector from engineering and hydrology, the water market managers experienced a startling inversion. They found themselves managing not a river system and an irrigation system but a financial market. The main players, the ones who really drove prices and water allocation, were the banks and quasi-banks, many if not most of them from overseas. We could see this was a matter of concern and regret for the managers, and that they were going through a period of mental adjustment—shifting their thinking from natural resource management to financial management.

We noticed they spoke of banks in a particular way. The managers knew banks were important in the market, and that the policymakers had deliberately called on them to fulfil a specific goal. But the hydro-engineers were struggling to grasp that rationale, and to understand what impact the banks were having.

We also saw that the managers were holding on to two optimistic ideas. The first: there were alternatives to banks as the source of liquidity. The original designers of the water market had encountered several forks in the road. One was the decision between a regulated or deregulated market, and another was whether to use private or public liquidity. Policymakers had the option, for example, of seeking liquidity from a public body such as the Reserve Bank of Australia, or the Murray–Darling Basin Authority, which could stand in the market and provide liquidity at minimum cost.

The ultimate decision to use private liquidity was consistent with the choice to go down the deregulated-market track, and it was consistent with the spirit and ideology of policy circles at the time. In the 1990s and 2000s, the idea of a public solution clashed with the mood—in Australia as in the UK and the US—of deregulation, small government and relying on 'the market'.

But here is the optimistic thought. When the water market was designed, governments had looked to external parties for crucial liquidity, but things had changed in the intervening decade. Lessons had been learnt, and new ideas and tools had emerged. Maybe, Mr

X and Mr Y seemed to think, the policymakers might change their minds about the source of liquidity.

The other optimistic idea: the liquidity role of external parties might have a time limit. In principle, Mr X said, once the water market had matured, there might be no need for any dedicated source of liquidity—public or private—because the market itself would be sufficiently liquid thanks to the normal, day-to-day trading of farmers according to their water needs.

We felt this idea was especially important to the managers—the idea that, at least in theory, the bankers' role in the market could be phased out, and the market would revert to being more about hydrology and irrigation, and less about investment banks and hedge funds.

'The banks have changed everything,' Mr Y said. 'That is where you must look to understand this market.'

CHAPTER 12
DARK MARKETS AND DARK POOLS

Ostensibly, banks and quasi-banks were allowed to trade in Australian water—even if they owned no land and had no interest in agriculture—because their presence would provide liquidity. But the institutions that rushed into Australian water were a far cry from the popular idea of 'banks'. Between the 1980s and the 2000s, the essence of banking and financial markets had been transformed more than once. To better understand the non-farming water market players, we had to grasp the nature of 21st-century finance, and what the big banks had become in the era of deregulation and FinTech.

The libertarian economic ideas that influenced water policy in the 1980s and 1990s also penetrated stock markets and the trading rooms of banks. Until the 1980s, the London Stock Exchange was notable for its rigid, time-honoured traditions such as minimum commissions and the separation of 'jobbers' from brokers. Several of the rules were intended to manage potential conflicts of interest, such as between brokers' incentives to serve their clients and the incentive to serve themselves.

The traditions attracted the attention of Thatcher-era competition regulators, who judged the rules to be 'restrictive practices'. That judgment spelled potential doom for the financial leaders of the City of London. But the Stock Exchange and the UK Office of Fair Trading reached a compromise. The office would drop legal action under the competition laws if the exchange committed to reforming itself by the end of 1986. The result was London's 'Big Bang', a radical softening of the laws and restrictions that had long governed British banking and broking.

The Big Bang reforms killed many sacred cows. They allowed, for example, outside ownership of British stockbrokers. American and Continental banks and brokerages would soon dominate British finance. The Bang reverberated far and wide. A wave of banking liberalisation spread around the world, and especially throughout the Anglosphere.

In the US, banks jumped state borders and other intra-national and international barriers. In Australia, the awarding of banking licences to foreign institutions was a cornerstone of financial deregulation. Japanese, European and especially American banks set up around the world and became truly global concerns. Cross-border financial flows exploded and the shape of finance changed forever.

Deregulation had been a deliberate decision of governments, but the world stumbled on the next major financial transformation almost by accident.

A traditional and unglamorous form of lending, residential mortgages were a staple product for many commercial banks and all of America's savings and loan (S&L) institutions, which are analogous to Australia's building societies. A mortgage is a simple, low-return instrument, or is it? Was there some hidden complexity that could be unlocked?

In the early 1980s, the sleepy S&Ls were in trouble due to light oversight and heavy fraud. At the height of the S&L crisis, these non-bank financial institutions were desperate. To stay afloat, they needed to offload assets—even the boilerplate mortgages they'd originated and tended on their balance sheets.

For Lewis 'Lou' Ranieri of Wall Street investment bank Salomon Brothers, this was a life-changing opportunity, and a market-changing one. He and his team of dealers (the same guys who trained Enron's energy traders) bought thousands of S&L mortgages cheaply, bundled them up, then sliced the bundles to create novel assets, or securities, to sell. This process, too, has an ear-bothering name: 'securitisation'.

Income from the new securities depended on interest payments on the underlying mortgages. From the point of view of investors, the resulting products looked a lot like government and corporate bonds, and they were certainly marketed that way. But under their glossy, safe-looking veneer, the new securities were heavily engineered with sophisticated maths and other methods, so that risky, lower-rated mortgages could form the basis for highly rated bonds.

Traditional banking had been about banks writing loans and then keeping them on their books, to be husbanded like crops. But now, under the banner of securitisation, banks rushed to disassemble mortgages and to shift them off their balance sheets—a process known as 'originate to distribute'. Between the early 1980s and the eve of the 2008 global financial crisis, the total value of mortgage securities backed by loans from government-sponsored lenders such as Fannie Mae and Freddie Mac grew by an amazing 2 million per cent, from US$200 million to US$4 trillion.

The originate-to-distribute model of residential mortgages largely replaced traditional home lending. Securitisation gave banks access to an enormous pool of global savings with which to originate more mortgages. It also gave them a bewitching taste of financial innovation and fast dealing. For the banks, there was something wonderful and mysterious in the creation of mortgage-backed securities. Making shiny, new, tradable assets from boring old home loans seemed like a magic trick. And the banks would perform the same trick in other places.

Deregulation and securitisation were intertwined with other trends that broke down boundaries between different markets and different types of institutions. In 1998, for example, Wall Street behemoth Citibank merged with the insurance company Travelers, which had acquired the investment bank Salomon Brothers. This instance of 'bancassurance' forced the repeal the following year of the Depression-era Glass–Steagall Act, a step that allowed ordinary commercial banks to move into high-risk, high-octane investment banking. (The act had previously separated the two types of banks, in part to avoid conflicts of interest.)

The newly deregulated world fuelled a red-hot boom in trading and financial innovation, and it created a smorgasbord of opportunities for banks, which embraced other new products and business lines besides securitised mortgages. Derivatives and swaps, for example, created new profit lines, especially in the 'over-the-counter' (OTC) market and 'dark pools' (private trading environments available only to financial institutions).

Tradable options are called 'derivatives' because their value is derived from an underlying asset, such as a stock. A 'swap' is a contract to exchange one set of payments or cashflows for another. An example is an agreement to swap the stream of dividends from a parcel of stocks for the interest payments on a loan.

The idea of an over-the-counter market sounds bland and benign, like over-the-counter drugs. It smells wholesomely retail, but it is actually wholesale—and neither bland nor benign. 'Under the counter' would be a better description. In the OTC market, sales of derivatives were negotiated privately and bilaterally—usually between a bank and another party—without going through a central market exchange or some other type of intermediary. A thick aura of mystery surrounded this sphere of banking and deal-making.

Banks controlled the OTC derivative contract terms, and the OTC market as a whole, just as they controlled the market for securitised mortgages. The banks could create their own rules, and they could levy fees for 'making the market'. They could build in margins (OTC derivatives have wider 'bid–ask' spreads than exchange-traded ones). They could stand between parties as the ticket-clipping intermediary. And they could set prices for the products they created. For all these reasons, banks sprinted into the market for derivatives.

Customers, too, jumped into derivatives and swaps, mainly because these hybrid securities allowed companies and wealthy people to hide income, reduce their tax, and otherwise escape government rules and imposts. Sales of derivatives exploded, and the majority were soon being assembled and sold in the bespoke, bilateral, off-exchange way. For banks, therefore, the OTC market proved to be extremely profitable. And because the banks set OTC derivative prices and otherwise controlled the market, all sorts of behaviours could occur that would normally be impossible, or illegal, on open exchanges.

In the OTC 'dark markets', for example, the banks could actively 'trade their own book' (which means trading as a principal, on their own account and at their own risk) rather than act as a representative for customers and clients—why trade for pesky customers when banks could enter the market by and for themselves, or better still, use customers' information against them for extra profits? In the OTC markets, banks would come up with a raft of ingenious ways to make money in the dark.

The new financial products wrought an important change in capitalism. Investors chasing high financial returns were no longer interested in 'buy and hold' strategies. Instead, they sought profits from active, intra-day trading—and volatility became hugely valuable. Volatility in financial markets refers to the extent to which, over any given period, prices move up and down. Thanks to financial innovation, volatility became a principal driver of banks' profits.

By actively buying and selling, and by using derivatives, banks earned vastly more than they could from merely holding an income-generating asset. At any time of any day, in a rising or a falling market, banks could make money, provided securities prices moved. In this faster, more cutthroat financial world, up was up, down was up, all times were good times, and the game could go on forever in a way that was largely invisible to outsiders.

Allowing major commercial banks and investment banks to make money in this way was a wrong turn for capitalism. It altered their incentives in ways that would ultimately separate them from their traditional, useful place in the economy. They could make a lot of money with little or no risk, and operated with hardly any scrutiny from regulators, who often had little idea of what was really going on, and who relied on abstract 'market forces' to ensure the banks behaved well.

—

The big banks did, however, get into legal trouble from time to time.

Three of Australia's Big Four banks were caught rigging the bank bill swap rate between 2010 and 2012. The scheme was clever and highly profitable: the swap rate is a benchmark rate used to price trillions of dollars of assets. Small changes in the rate therefore had big and lucrative consequences. In other cases, too, major local and foreign banks were caught out doing all sorts of unedifying things.

In 2019, Australian legal action was launched against five investment banks who allegedly colluded to rig foreign exchange rates between 2008 and 2013, again at the expense of businesses and investors. The traders reportedly colluded via internet chatrooms, which were given names such as 'A CoOperative', 'The Cartel' and 'The Mafia'. In a similar US case, some of the same banks were caught up in another rate-rigging scandal in which they had to pay a fine of more than US$2.8 billion.

In another recent case, US federal prosecutors accused senior J.P. Morgan precious metals traders of conspiring to manipulate markets and defraud customers, such as by placing fake orders that created a false picture of prices and demand. Similar accusations were made in relation to manipulation of the market for bond futures. In a multi-agency settlement, the bank admitted to wrongdoing and agreed to pay more than US$920 million in penalties and restitution. This was J.P. Morgan's fourth criminal case since 2012.

Other financial market participants, too, were caught using sophisticated strategies to manipulate markets. One celebrated example involved using super-fast digital connections and 'high-frequency trading' to get microseconds ahead of buy orders in the stock market (a practice described by Michael Lewis in his book

Flash Boys). Investment banks, hedge funds and other trading outfits used this tactic to make billions of dollars by on-selling the shares at an incrementally higher price.

As Australia's financial regulator ASIC noted in 2015, the global financial crisis—particularly the collapse of investment banking group Lehman Brothers, and the threatened collapse of US insurance group AIG—demonstrated the magnitude of the new risks in finance, especially those associated with OTC markets and new derivative products.

What did it mean to be a bank in the 21st century? The safe, conservative, 'hello neighbour' era of branch banking was long gone. When it came to Wall Street today, the common, familiar idea of what constituted a bank was simply no longer relevant.

The giant, money-centre institutions that had conquered global finance were agile, high-tech, predatory enterprises, making money and taking risks in the dark, at the expense of customers and also of taxpayers, who unwittingly gave the banks an unlimited gambling licence in the form of a generous public guarantee. The banks had made huge investments in advanced predictive models, trading software, exotic securities, super-fast computers, quantitative geniuses and automated 'bots'. They were always 20 steps ahead of the governments that sought to regulate them.

This, we thought ominously, is who the Basin governments had let into the water market.

CHAPTER 13

THE SMARTEST GUYS IN THE ROOM

The investment bankers and hedge fund managers who quickly entered the Australian water market did so in multiple forms and guises. They set up special-purpose vehicles. They entered partnerships, joint ventures and agency relationships with local players. They bought stakes in large farms and agribusinesses. Some bankers even left their Wall Street employers altogether to set up shop on their own account in the new water world.

The bankers and other traders seem to have quickly understood the scale of the opportunity, and the way to seize it. Once in the market, they saw that they needed to invest in specific expertise, and to assemble dedicated, water-focused teams. They set up trading rooms and populated them with dealers, modellers, accountants and other professionals.

'It's an incredibly complex market,' ACCC deputy chair Mick

Keogh said in 2020, 'and major players put considerable resources into understanding it. They have up to 20 analysts looking for opportunities. That is the level of brain power you need to operate powerfully in the market.'

'All of a sudden,' water broker Desiree Button said, 'we've got banks and funds that have employed people who do the trading for their living. They've got university degrees and they're seeing what's going on.'

'When the government opened up water trading,' farmer Chris Brooks said, 'we suddenly had rooms full of these young smartarses.'

Using glimpses and hints and snippets from multiple sources—such as brokers, traders, officials, regulatory reports, rural media, talkative fund managers and the water market underground—we've pieced together portraits of the traders themselves, plus a sense of life inside the trading rooms.

'A few [of the traders] are older, red-faced, "Sir Lunchalot" types,' a market insider told us. 'Heroes of the 1980s trading room floor. Refugees from the days before Fringe Benefits Tax. But most are young turks.'

Many of those turks and Chris Brooks' smartarses are as gung-ho as Wall Street mortgage dealers or Houston-based energy traders.

'To picture the culture,' another insider said, 'just imagine a group of teenagers making prank phone calls—while sitting on $100 million of play money.'

Traders were plucked from all directions, creating a trading room culture that blended IT, hydrology, farming and high-stakes gambling. Some had worked for banks overseas; some were mentored by dealers and financiers from the major financial centres; and a lot were local recruits: former estate agents, former farmers, rural bankers, country accountants, engineers, physiotherapists. A surprisingly high proportion were literally used car salesmen.

For personnel in the foreign-owned trading rooms, the arrival of visitors from 'head office' was always an occasion for learning, trepidation and excitement. For the visitors, the experience was fascinating but jarring. On Wall Street, water traders looked exactly the same as the bond traders and swap dealers. But in Australia, professional water traders had an identifiable, in-between persona that came from the odd juxtaposition of Wall Street methods and rural locations such as Wagga Wagga, Mildura and Deniliquin. The persona was Wall Street but also farm gate.

There was the occasional active trader who was an unapologetic, Porsche-driving spiv, flagrantly chasing easy money—and telling everyone about it. At rare public meetings of investment funds, there were moments when high-profile water barons couldn't contain their delight. They just had to share graphic stories about the profits to be made in the market. Some traders even appeared on television to brag about their wins.

But most traders maintained a low profile. More likely to drive a HiLux or a LandCruiser, they kept their success quiet, in part because of the rural ethos of not crowing about money, and in part because the more aggressive traders and brokers feared they would be targets of violence from people on the losing end of the most excoriating deals. Punch-ups at pubs and on farms were not unheard of.

There is another reason, too, why the smarter companies and traders mostly stayed quiet. Opportunities from trading Australian water were appreciated and celebrated in Melbourne, Sydney, Frankfurt, Shanghai, Chicago, Boston and New York. But the big-money banks didn't publicise the sweetest deals. Instead, they kept them for themselves, and traded them, year in, year out, on their own account. They did this very quietly, so their customers didn't know, and the regulators didn't turn off the dark money tap.

For the traders, this fact alone—that much of what they did by necessity took place in secret—was itself an attraction and a source of excitement, as was the daily trade in rumours and scandals and gossip. These perquisites added to the general feeling that trading water at high speed and for big rewards was an adrenaline rush.

Farmers didn't know who was on the other side of their trades—and when they wanted to find out, typically they could not. The traders maintained their cloak of secrecy, and the water market regulatory requirements were little help in providing transparency. There were few requirements on banks and other traders to disclose or account for their actions in the market. As Mick Keogh of the ACCC said, active traders could 'transact millions of dollars of water without an ABN [Australian Business Number] or tax file number'. In some states, there were no water registers at all, and in others the registers were incomplete and hard to access.

Farmer Laurie Beer told us, 'There's people who say, "Oh, yeah, there's a register out there where you can look it up." Anyone who has seriously tried, it's the same story. You can't really identify exactly who it is, unless you've got some pretty serious statistics on them already. You're really feeling around in the dark to try and work out who owns what.'

'I don't know why there's so much concern around having a water register,' broker Desiree Button said. 'The only reason you would be sceptical and opposed to that, is if you are doing something not quite right or there's a conflict in terms of potentially foreign investment trigger points. Even the overseas investors, you can't nail down who they are.

'I think it's in Australia's interests to know who owns all the water. You might see the name of a bank or a fund but there are certainly other specific water holders, investors and shareholders that could be the majority stakeholders in it, here and overseas.' She added that it was the volumes of water, more so than the monetary value, that should be transparent, 'because we might start seeing some trends in that ownership and where all that water potentially sits. That would be important to know.'

'A lot of investor money has piled into water,' farmer Ben Dal Broi said. 'From where, I don't know. For all we know it could be Brisbane, Beijing, Bangkok or Brussels.'

'The banks just kind of arrived,' a local trader said. 'Almost overnight. It felt like they were just taking a bit of a punt. At least at first.'

In part due to the cloak of secrecy, and in part because the definition of 'trader' had fuzzy edges, it was hard to tell precisely which banks and other outside investors had decided to play the Australian water market in a big way, and it was hard to tell which individual traders were active in the market on any given day. Some active water trading blended into other trading-related activities of organisations such as super funds and agribusinesses, for example, and some highly active participants in the market were brokers, or especially switched-on farmers, rather than professional traders.

But based on what we've seen and heard, we think the dedicated water trading rooms numbered at least in the dozens; and of these, perhaps a handful of the teams really knew what they were doing. That handful of trading rooms, which made the running for all the others, were backed by the smartest banks and the strongest modellers and programmers. And they had the best information about Basin farmers.

—

Information is key to all markets. It drives the behaviour of participants, which in turn drives prices. Information is a doorway to profit.

Water prices in particular are highly sensitive to information. For example, NSW Water's 15 October 2020 announcement of a 9 per cent increase in irrigators' water allocation caused prices to drop by around 15 per cent on the MIL exchange. A single fall of rain can also cause a 15 per cent price movement, or an even larger one. For water market participants, big money is to be made from getting ahead of such swings in prices.

Other price-sensitive information included the details of water allocations, decisions of water authorities and agencies to buy and sell water (including large environmental purchases), and inter-valley transfer openings. In addition to watching those decisions and openings intensely, the banks and their traders studied the market rules—all the plans and templates and guidelines. They collected general market knowledge, and specific information about conditions on the ground. The most valuable information was about farmers' water holdings and plans and needs.

'It is fascinating,' Desiree Button said, 'how certain large players seem to have an amazing ability to know everyone's water licence, how much water is on them, how much water at any given time is available for allocation throughout the year. It's amazing for all the clients around here to get rung up out of the blue by the large traders. How they've been able to get all that private and confidential information is quite extraordinary.'

The traders gathered all sorts of intelligence about farmers. They learned how crops differed in their productive life. Almond trees, for example, are reliably productive for around 20 years, whereas pistachio trees keep producing for between 60 and 300 years. The traders also discovered that different types of farmers bought and sold water in predictable ways—such as in response to price movements or rain or other identifiable factors—and often moved seasonally as a single group, like a herd. There were predictable peaks in cotton farmers' demand for water, for example, and there were predictable price spikes at the height of summer.

'There can be a lot of speculation and emotion driving the market, too,' said Ben Dal Broi, 'especially when there is talk of a heatwave.'

Tom Rooney, founder and director of the Waterfind exchange, told the ABC in November 2019, 'We are about to hit the peak of the water demand in late November through to about February, where we've got the highest demand for water use in the system. And sometimes the highest amount of water use will also mean more demand [reflected] in the temporary water market price.'

The river system managers, too, were predictable in their choices and actions that influenced the value of water—such as water management decisions around the Barmah Choke. Understanding these decisions, and predicting them, became a daily concern for the traders.

'The traders got better and better,' Desiree said. 'Smarter and smarter.'

This investment in expertise allowed the traders to find the best wrinkles in the market and the best trading opportunities: lucrative combinations of times and scales and types of water rights.

The traders used the tools of statistics, IT, hydrology and behavioural economics. They built complex market models with long equations that factored in uncertainty as well as variables like weather, river flows and government decisions. These models allowed the traders to capture and predict seasonal conditions and farmers' seasonal behaviours. The traders used this knowledge directly to inform their transactions in the market. They knew when to buy and when to sell.

The traders had the resources to collect and compile rich libraries of highly specific information. They could also pay brokers and other market participants to share valuable facts. But smaller farmers and even groups of irrigators couldn't afford to make the same investments in whole-of-market expertise and understanding the rules. Nor could they afford the big capital costs and transaction costs associated with intensive trading and seeking market advantages.

There was an 'information arms race' in Australian water, but it was extremely one-sided. The traders came to know a lot about the farmers, but the farmers knew next to nothing about the traders. Most of the time, the farmers were like Laurie Beer—they didn't know who was on the other side of their trades. This is a classic case of information asymmetry, which in turn is a textbook cause of market failure.

CHAPTER 14

BUY LOW, SELL HIGH

Thanks to financial market insiders, we now knew a lot about the transformation in global finance, and how the big banks made their money. But how relevant were those tactics to Australian water? Were big banks and quasi-banks really playing the market for profit, or were they merely providing 'liquidity' and helping to allocate water efficiently, just as the market designers had hoped? What was happening on the other side of Laurie Beer's Harvey Norman laptop?

Arbitrage is a classic trading strategy and it has classical simplicity: buy low and sell high. It can take countless forms, some of them very familiar, like buying and selling under-priced and misdescribed collectables in country-town swap meets, or in the *Trading Post*, or on eBay. Legendary examples of buying low and selling high include 'barn find' discoveries of pre-war cars and motorcycles. The television series *American Pickers* is all about that kind of arbitrage.

In financial markets around the world there is active arbitrage trading every day—such as between stocks listed on multiple exchanges, and between agricultural commodities sold in different markets. In the OTC derivatives market, major banks enter complex contracts that are essentially about arbitrage between different valuations of risk, or different views about the future. For banks and hedge funds, extreme and intricate forms of arbitrage are bread and butter.

The famous fund Long-Term Capital Management (LTCM) is a spectacular example. The fund's traders scoured markets around the world—such as those for securitised mortgages, government bonds and foreign currency—in pursuit of market discrepancies. The idea was to hunt down instances where the same financial asset was priced differently in different markets or sub-markets. Sometimes the price differences were minuscule, but with enough leverage and capital, even minuscule differences were a route to huge profits.

Some of the best minds in academia and Wall Street were behind LTCM. It was created when leading economists Robert Merton and Myron Scholes teamed up with ex-Harvard professor Eric Rosenfeld; John Meriwether, former head of bond arbitrage at Salomon Brothers (who made his first fortune during the S&L crisis); David Mullins, who had also worked at Salomon Brothers as well as the US Federal Reserve; and a hand-picked band of other Wall Street traders.

LTCM's investors were mainly Wall Street banks, who each joined the fund with a minimum investment of $10 million, and with expectations of hedge fund–level returns. Wall Street firm Bear Stearns executed the LTCM trades. Rooms of quantitative analysts structured the LTCM positions, which numbered in the thousands. As Niall Ferguson wrote in his book *The Ascent of Money*:

> In October 1997, as if to prove that LTCM really was the ultimate Brains Trust, Merton and Scholes were awarded the Nobel Prize in economics. So self-confident were they and their partners that on 31 December 1997 they returned $2.7 billion to outside investors (strongly implying that they would much rather focus on investing their own money). It seemed as if intellect had triumphed over intuition, rocket science over risk-taking. Equipped with their magic black box, the partners at LTCM seemed poised to make fortunes beyond even George Soros's wildest dreams.

Through entities such as LTCM, Wall Street banks turned complex arbitrage into a fine art.

In the Australian water market, we found, there were many different modes of short-term and long-term arbitrage: arbitrage between exchanges; between on-exchange and off-exchange trading; between different types of rights; geographical arbitrage across farms, valleys and states; and intertemporal arbitrage across minutes, hours, days, weeks, months, seasons and even (thanks to carryover) years.

Waterfind director Tom Rooney told the ABC in 2019:

> We saw recently, in the last week for instance, the Murrumbidgee market open up with a 15,000-megalitre [inter-valley transfer] opening, which means that growers in the Murrumbidgee could actually transport water out of the Murrumbidgee into the Murray system, and we saw that close just as quickly as it opened. And that was really driven by Murrumbidgee growers wanting to get their water out of the Murrumbidgee

> into the Murray system to maximise the value on their temporary water price, because the temporary water difference might have been $200 a megalitre and that additional value that has been shifted out might have meant another $3 million to water holders that got their water out.

As the foreign bankers had noticed right from the beginning, there was a lot of intra-day volatility in the water market—and this meant traders could make money on any day and at any price level.

The water traders were engaged in a permanent hunt for wrinkles in the market. They sought anomalies that would allow them to buy under-priced water and sell it at a premium. The wrinkles included short-term differences in water prices between different regions and different water exchanges. Particularly in the first decades of Australia's water markets, there were many opportunities to make low-risk arbitrage profits, especially through inter-region and inter-exchange trading.

Specific modes of intra-day arbitrage included 'natural arbitrage' (pursuing existing, in-market opportunities to buy low and sell high), 'forced arbitrage' (making strategic purchases, usually backed by large capital, to ramp up the water price and then sell at the higher price level) and 'false arbitrage' (more about that later).

Natural arbitrage involves buying under-priced water rights and then immediately selling them back into the market. This type of arbitrage depends on information and speed. Forced arbitrage works in a different way. It depends on heft. Water broker Clinton Davis called it 'a type of market manipulation'.

'Anyone with a lazy million bucks could easily manipulate the market,' Clinton said. 'Just start buying water and then drop the guts out of it.'

Another trading strategy exploited anomalies between the current and future value of water rights.

'Traders make bets about future prices,' Jerilderie farmer Peter Burke said. 'They can entice a farmer into a two- or three-year deal for water—locking in the price—because they know they have better information about where prices will go in the future.'

This is similar to the gambles that banks regularly make between the pricing of fixed-rate and variable-rate mortgages. These gambles almost always turn out in the banks' favour, because they have better information about the price of credit, future market conditions, and the borrowers' willingness to pay.

The same intertemporal wrinkles could be exploited in another, more complex way. Using sophisticated algorithms and clever mathematics that were transplanted directly from Wall Street dealing desks, the water traders could reap arbitrage profits by trading between the ordinary 'spot' market and the 'forward' market for contracts for future delivery. (They could exploit price differences, for example, between multi-year water leases and equivalent bundles of forward contracts.) This was just one of the numerous kinds of arbitrage of which farmers and even regulators had very little inkling.

Decisions of public agencies and water managers created some of the most significant and lucrative arbitrage opportunities.

'The Choke,' farmer Chris Brooks said, 'is the worst example I can give you of a water trading distortion. There are two different zones: zone 11 downstream, and zone 10 upstream. You cannot buy water from here and sell it downstream. That's why there's two or three hundred dollars' difference in the price of water upstream as opposed to downstream.

'When you buy water from farmers here and farmers there, the water has to maintain its original status, meaning it is either zone 10 or zone 11 water. So if it's zone 10 water, it cannot be taken downstream of the Choke. But we have rock-solid evidence that a public body deliberately transferred 500,000 megalitres of water downstream in a drought year and traded it to the higher value markets, and didn't think we would find out.

'Just to add to it, to the type of examples of this stupidity, the bigger corporations have this insane ability of knowing the markets and knowing the way around all of the bloody rules and regulations. They would buy a chunk of water down in zone 11. And knowing the rules as they do, they would come up to zone 10 and, let's say, buy 20,000 megalitres of water from 20 different farmers, a thousand a time. They can't trade it downstream until water comes upstream.

'You tell me what commercial enterprise would deliberately transfer 20,000 megs of water from one zone to another where the price is $200 less. But they do it because it creates a credit. They'll automatically transfer the 20,000 megs to MIL, and then they'll perform an inter-valley transfer. They'll move this water out of the Murray irrigation system to the Murrumbidgee or the Goulburn. And both of those outlets run back into the Murray downstream of the Choke. And so they end up with their water down there, and some very valuable credits, too. And you've got irrigation schemes actively facilitating these transfers.'

Arbitrageurs can only survive if, at any given time, there are different market prices driven by different understandings of what water is worth. As a result of how information is dispersed in the water market, there are many opportunities for information-based arbitrage across different regions and market segments. Price-sensitive information is spread across multiple exchanges and sub-markets and brokers. A big price movement on one exchange will not be immediately reflected on another. This is a critical point about the water market: at no particular moment is there a single, universal 'price'.

While the amount and the quality of the information was important to the traders, they also cared very much about time and speed. Information becomes stale very quickly. Water traders made sure they had comprehensive, real-time information, whereas many farmers didn't have an up-to-date picture, or a complete one.

Big swings in prices mean big swings in expectations of value—but those expectations were frequently out of synch across different exchanges, thus creating opportunities in the market. For example, in 2020, prices of temporary water fell from around $800 or $900 to around $120 or $150 a megalitre. When the market found its new level, traders and brokers could easily sell water to farmers at a much higher price, because farmers had out-of-date information about what the water was worth.

The flipside was also true. When the market shot up to $800 a meg, it was easy for a broker or trader to induce farmers to sell at $700, because the price was recently $600 so the farmers still felt as though they were getting a good deal—when in fact they were being screwed.

For arbitrageurs, differences in perception were a pathway to big returns. If farmers were prepared to pay $150 per megalitre when the market was really at $100, that sounds like a small difference—and we heard many examples of trades like this—but in fact it delivers a huge return for the seller. This is an important reason why confusion and dysfunction in the water market are sure-fire routes to guaranteed profits for well-informed traders. Those traders sometimes complain about the market's immaturity and fragility, but in reality they relish it, and thrive in it.

Unlike the farmers, the traders watched all the exchanges at once, and could exploit fleeting differences in prices and information across exchanges. The traders' proprietary systems collected and consolidated information from multiple exchanges and multiple sub-markets—so they could learn from it and trade on it. The private systems had better and deeper information than even the individual water exchanges. This comprehensive picture was extremely important for traders and brokers looking to profit from day-to-day trading. For the most well-informed and sophisticated traders, holistic information was a means of market control—potentially more useful than owning and running the exchanges themselves.

'Investors look at those occasions,' Ben Dal Broi said, 'when it's possible to sell water on the temporary market for $850, like in early 2020. And at $8,000 for a megalitre of high-security entitlement, you go, "Well, that's around a 10 per cent return. It's fantastic and one would do that all day." But if the water price comes back to $150 per megalitre like it did today [January 2021] and you bought it for $8,000, you go, "Oh, this has got hairs on it."'

'Water trading is a lot like the stock market,' Peter Burke said. 'It's a gamble. Prices can move up or down overnight by $200 a meg. That's a movement of more than 20 per cent in 24 hours.' (In the early years of water trading, volatility was even higher. Price movements of 1,000 to 2,000 per cent within a year were common.)

'When you look at it,' Peter said, 'last year and the year before it was up at $900 a meg. Right now it's $150. It got as low as $120 not long ago.'

We thought Peter was right to liken the water market to the stock market. They both have elements of gambling. But even with regular 20 per cent movements in prices, this is an incredible level of volatility: an order of magnitude higher than the stock market's daily movements. Translated to stocks, it would be like suffering a Black Monday crash every second day. No one in the stock market would tolerate such a rollercoaster.

There are other differences, too, between water and stocks. Unlike in the stock market, few people in the Australian water market see in real time the full profile of bids and asks that are 'out of the money'—meaning the offers that are below the best offer, and those asks that are above the lowest asking price. Farmers in particular typically have no clear idea of the 'depth' of the market or the precise market forces that are driving water prices at any given moment.

One anonymous broker offered an explanation for why the full set of bids and asks was not published.

'I suspect they don't publish it because it would be terrible!' she said.

The profile, in short, would show farmers were being ripped off, and the private providers of 'liquidity' were not doing their job. The bid–ask spread provided further evidence that the liquidity providers had failed. In a liquid market, the spread should be narrow, but in water it is perilously wide.

Argyle Capital Partners is a major water owner in the Basin. The company made the following observation in an October 2020 submission to the ACCC.

> In almost all daily observations, and by comparison to most other markets, the spread between bids and offers is wide. This is not a failing of the transparency of the markets, rather it is an indication of the illiquidity of the markets. As an example, yesterday (28 October 2020) one of the more prominent on-line broker

> exchanges for the market zone with the greatest annual turnover, Victorian Murray Zone 7 quoted buyers' bids from $180/ML and sellers' offers from $220/ML. That represented a $40/ML bid/offer spread; a margin 20 per cent above the bid. Consequently, depending on their willingness to cross that spread, two farmer irrigator[s] as buyers on the same day in the same market zone might get filled at very wide price differentials (purchased water allocation at $220/ML, or patiently waited for a seller to accept a bid at $180/ML).

Making things worse, the full set of bids and asks were dispersed across multiple platforms and exchanges, and some were not even online at all (such as those for off-market trades). This made it even more difficult for farmers to gauge the true market depth.

In the absence of universally visible prices and bids and offers, the traders could game the bids and offers posted by farmers on individual exchanges. Armed with their complete picture of all the different sub-markets, traders could use a range of tactics to their advantage: for example, they could place 'fishing' bids and offers in the intervals between farmers' bids and offers, so as to influence prices (in a way that made optimal use of the traders' capital) and to extract valuable information about farmers' willingness to pay.

Complex strategies such as these were aided by the traders' most valuable secret weapon—the water bot.

CHAPTER 15
WATER BOTS

Some traders worked from high floors in Melbourne office towers. (Compared to Sydney, Melbourne is much closer to the Murray—and even to most of the Murrumbidgee.) Some sought foreign citizenship and based themselves, ostensibly for legal and taxation purposes, in overseas havens. But the best traders found premises in the Basin itself, close to the water exchanges, and to the farmers and the rivers. Basin traders had a better understanding of the seasons and the conditions on the ground. And they enjoyed a few microseconds of extra speed—because their electrons didn't have to travel to Melbourne, Sydney, New York, London or Frankfurt.

For reasons that were widely discussed in the water market underground, the microsecond advantage was especially important. It complemented the use of Wall Street–style computerised trading and 'bots'. When we first heard the word 'bot' in relation to the

Australian water market, we were stunned. Not only had the banks transplanted modern financial arbitrage, they'd also brought their high-speed trading programs, one of the most striking manifestations of FinTech.

As late as the 1970s, the basic workings of the world's financial markets were substantially analogue and very clunky. At the New York Stock Exchange (NYSE), for example, ticker tape and pneumatic tubes were recent memories, and stock prices were expressed as fractions rather than computer-friendly decimals. (The NYSE did not go decimal until 2001.) Insurers at the famous Lloyds of London still relied on hard-copy actuarial tables.

In the 1980s, finance underwent a digital revolution. From around 1986, for example, bond markets and stock markets began introducing electronic placement of buy and sell orders. Electronic order matching soon followed, along with electronic settlement.

The emergence of electronic exchanges coincided with the proliferation of computerised trading strategies. Large brokers and investors used computer programs to submit buy and sell orders based on pre-set rules and ratios and market triggers. The programs used complex mathematics such as Markov chains and Brownian motion. Some algorithms were home-grown within banks, but others were bought from specialist programmers and mathematicians.

The first decades of electronic trading were full of disaster stories. Crude and stale program trades were a primary cause of the 1987 stock-market crash, and program trades were also central to the 2010 'flash crash'. But as computers and software became more sophisticated, so did the trading programs, and they claimed an ever-larger share of the market. By the 2010s, for example, the majority of share transactions on the NYSE and the Nasdaq were from computerised orders.

—

In many respects, speed was king in the water market—and this was a deliberate feature of the market design. The daily functioning of exchange trading and official processes worked on a 'first come, first served' basis. Examples included applications for inter-valley trade; and day-to-day, on-exchange, 'spot market' transactions.

Timing, therefore, was extremely important. In 2020, the Melbourne-based agricultural investor goFARM told the ACCC how speed dominated in the water market: 'the "fastest finger" wins when trading windows open'.

Water broker Desiree Button elaborated. 'The market is very liquid. Everything can happen in a split second, and people can make and lose, and manipulate, and make millions and millions of dollars in transactions that can happen in the stroke of a button.'

From our water network, and from computer-savvy traders, we learned a lot about the bots. Right at the beginning of water trading, the traders had rushed to increase their speed. They sank money into communications infrastructure. They hired programmers and quantitative analysts, and they imported program-trading software directly from other markets such as those for futures and derivatives. As a result, the 'fastest finger' was not a finger at all. It was a computer, operating at super-human speeds.

With the help of digital technologies, the water traders became a new breed of 'flash boys'. To play the water market, and to exploit wrinkles and loopholes, they used a variety of different bots and program trading strategies. The most common approach was to engage in a form of electronic arbitrage. Digital trades were automatically triggered when the computers detected bids and offers at specified levels; or when they measured price movements that exceeded pre-set thresholds; or when they calculated certain price ratios and differentials between valleys and other sub-markets.

Bots were especially useful for trades at idiosyncratic and unpredictable times, such as ones that involved moving water rights back and forth through the Barmah Choke. Having bots at the ready meant that the human traders did not have to be in the market or in official queues day and night, permanently alert. More importantly, the bots always won the race.

Water broker Clinton Davis told us about trading through the Choke. 'I'm amazed,' he said, 'by the ability of some brokers or individuals to be able to get water from above to below the Choke in any split second, when supposedly no one knows when it is going to open or not, when the volumes are available, because obviously DPI [the New South Wales Department of Primary Industries] say how many megs are available to go up above and from above to below, and sometimes that changes in a split second and all of a sudden that volume is gone. The same as inter-valley trading between the Murray and Murrumbidgee. It's taken up in milliseconds. I don't know how that's possible.'

Desiree Button encountered the same mystery. 'If someone from below transfers water up through the Choke, at the same volume you can go below. Sometimes it's thousands of megalitres, then it's available to go back down and for some unbelievable scenario that volume is automatically triggered. You have to lodge with WaterNSW and it's first in and best dressed. I know some people who seem to be able to get the water through every single time. I've tried 50 times but have only got it through once or twice.'

Desiree suspected bots were the explanation, and she was right.

'Any time the Choke opens for a small amount,' a trader confirmed, 'it's driven by bots.'

The Choke was naturally a focus for bot innovation. The

ability to move water reliably through this constraint, and to game inter-valley trading limits, was a crucial ingredient in the more significant and ambitious arbitrage strategies. It was the key that unlocked the whole market. The Choke was so important that it inspired its own bot race.

A Basin entrepreneur developed a program that gave paying users a priority position in the Choke queue. Not everyone could be first: the algorithm gave precedence to the highest bidder. Eventually the entrepreneur sold the algorithm, but other Choke bots emerged. Brokers, banks and quasi-banks have won the bot race by using faster computers and by deploying ever more audacious algorithms.

Other complexities and loopholes were focal points for digital innovation. There were arcane rules developed early on in the water market, such as 'grandfathering' of water, and these were now being used to exploit the market: traders worked out how to evade restrictions on inter-valley trading through clever, automated use of grandfathering and tagged water accounts. In their holistic, proprietary systems, the traders captured the complex rules in valuable strings of programming code.

'It's amazing we're even talking about this,' exchange manager Mr X said. 'Coding and bot trading in our water market. It's unbelievable. The market was supposed to be about irrigation. It was supposed to be about the water.'

The bot programmers could transplant techniques directly from other markets, but there were still challenges. The programs had to be customised to local conditions, with different inputs and triggers. Financial models had to be integrated with hydrological ones, and the resulting models had to connect with different water exchanges and official systems. This meant building electronic bridges between exchanges that were otherwise not connected or interoperable.

Regional centres such as Wagga Wagga, Deniliquin and Mildura had never before seen a programming effort of such complexity. The computers that hosted the bots were more sophisticated than anything at the local water corporations, the local banks, and even at the Bureau of Meteorology, which played a formal role in water market accounting.

'The traders have lots of computers,' farmer Peter Burke said. 'Financial models but also weather models, which they use to predict supply and demand, and the direction of the market.'

Model developers and bot programmers experimented with ideas and techniques from new and emerging sub-fields of finance and mathematics and computing: neural networks, Bayesian math, blockchain, fractal finance, big data and artificial intelligence (AI). Just as the water traders weaponised ideas from other markets, they also weaponised academic research.

Fractal maths had become a reality in finance. People searching for fractal patterns in commodities and securities markets said they'd found what they were looking for. Financial author Edgar Peters, for example, claimed to have proven US stock prices displayed 'memory': they were correlated statistically over four-year cycles. Benoit Mandelbrot found concurrences and correlations in daily, monthly and yearly price patterns in the market for cotton.

Similar harmonies and tautologies were identified in foreign exchange markets. Some financial crises, too, were found to be fractal. Efforts to discover fractal patterns in the water market would later bear fruit in unexpected ways.

—

According to a Renaissance proverb, 'There are sparrows and Florentines all over the world.' In journeying from Wall Street to Wagga Wagga, market bots had migrated further than the bar-tailed godwit. Not only were program trades present in Australian water, they were becoming more sophisticated.

Traders experimented to find out which digital strategies worked best, and they favoured exchanges that were better suited to their bottish strategies. They found that the bots and program trades were extremely useful in searching for arbitrage opportunities, and this was a fact of great interest in the trading rooms.

Both inside and outside the market, the most successful traders watched their competitors carefully. Banks and hedge funds could tell what kinds of automated trading were taking place in the water market, and the techniques of rival traders were valuable data. Traders borrowed and stole ideas from each other, and from dealers in other financial sectors. The water bots and their wranglers were learning.

Aided by their exceptional speed and extensive market information, the bots thrived. We learned they were behind mysterious trades in small quantities of water, some of them executed as transfers between accounts. Traders also programmed the bots to make 'fishing' bids and offers. These allowed the traders to pick up underpriced water instantaneously, and to extract information about the imminent decisions of buyers and sellers.

One automated strategy involved 'scraping'. Traders exploited the process of applications for inter-valley trade to 'scrape' data that revealed when an inter-valley limit would open. Traders could also spam the trade process by submitting multiple applications for different volumes that maximised their chances of success.

'No question that well-placed market participants are employing these strategies,' a Basin businessman told us, 'via their IT systems and access to market metadata.'

The use of these tactics and technologies gave professional traders a permanent advantage. They were always fastest. They would always be able to snaffle up the best choice of the available buys, and they would always be first in the queue to sell at the most opportune times. (Some traders bragged they could deal in complex financial products across multiple exchanges within a hundred-billionth of a second.)

'The big water traders, who have dedicated teams, are now competing for that lesser volume of water,' farmer Shelley Scoullar said, 'now that there is even less available. And your small family farmers are just competed out of the market.'

'I wouldn't know about this year,' farmer Laurie Beer said in September 2020, 'but I reckon last year in particular and probably the year before, when we were on zero allocation, water was very scarce. It just seemed to be a remarkable coincidence that there always seemed to be just a little bit of water on the market all the time, and it was around about the same price. It said to me that someone was actually doctoring it. Just putting enough on to serve their purposes.'

Who was on the other side of these static 100-megalitre offers that Laurie saw in the market? A computer program. The strange offers he saw on his screen were bots fishing for money. The standing offers served multiple purposes. They signalled prices. They were a way to pick up profitable bids. And they were a way to extract information about farmers' trading plans and 'willingness to pay' at specific prices and quantities.

The IT tactics evolved in extreme directions. Traders learned how to use computerised attacks to 'break the market'. They could stop and start trading on particular exchanges and in particular sub-markets. One way to do this involved another form of spamming: traders could bombard an electronic market platform with computerised instructions and junk information in order to clog it up and close it down. For the trading room turks and quants who'd grown up playing Warcraft and Warhammer, nothing was more exciting than a full-force bot invasion.

Stopping trading even for a few minutes could be highly useful if done at well-chosen times. Techniques such as these allowed traders to channel trading traffic and to lock out rival participants, such as those trading on different exchanges, and therefore to reap a price premium. In addition to the information arms race, therefore, there was a digital war against farmers, irrigators and water exchanges. That war, too, was entirely one-sided.

As part of their cloak of secrecy, the traders adopted the names of celebrities and porn stars as aliases and code-names. And they gave names, too, to their money-skimming, spam-attacking, market-dominating bots. In everyday conversations, the fastest computers and most profitable algorithms were referred to affectionately and reverently with names like 'The Monster', 'Goliath', 'Fatman' and 'Big Betty', who used to dwell in banks of servers in a Deniliquin office, but then moved to a home somewhere-nowhere in the cloud.

Apart from appreciating the power and elegance of the coding and the maths, the IT-savvy traders noticed resonances between the digital world and the water world, such as how water was measured in mega, giga and terra units—just like bytes.

The farmers we interviewed had noticed the fortunes amassed by the traders.

'I've seen the transition,' Riverina farmer Lachlan Marshall said, 'from when water was tied to land to when it was separated. What it inadvertently created was a money-making mechanism for people who have no interest in agriculture whatsoever. I describe it as the Wild West. There are huge amounts of money to be made by people who have the power, the flexibility, the capital to do so. People will argue farmers have that ability as well. And yes we do, but our business models are built around producing food or fibre, not being traders.'

In 2019, Basin farmer Rob McGavin estimated that, in the preceding year alone, his fellow farmers and irrigators had had to pay up to $400 million more for their water due to the impact of non-farming traders and speculators. On average, that figure corresponded to a 30 per cent margin on all water trading in the Basin. In extracting that margin, the water traders and speculators accomplished something few other people in financial markets had been able to pull off. They'd achieved consistently high returns while taking on very little risk. Some were able to retire in their 30s or 40s and moved to Byron Bay or Sunshine Beach to enjoy their water profits.

The money they made was a straight loss for Australian agriculture and rural communities. 'This is money we would otherwise reinvest in strengthening our businesses,' farmer Ben Dal Broi said, 'and supporting our rural communities.' Some traders and brokers were morally inured to the impact of that loss on Basin communities. Others, though, had qualms about it, especially those who lived in regions where farms were struggling or closing. Nevertheless, they continued to trade.

Water broker Clinton Davis confided that he had mixed feelings about his trading activities, but he was reconciled to the amount of money he was making in the water market. 'I've got very loyal clients,' he said. 'Is my business as big as some? No, but I turn over millions of dollars of water, both temporary and permanent. I make a very nice living out of it.'

Clinton's fellow broker Desiree Button also did very well. 'The sad thing about it,' she said, 'is that I take the view—and it's a very contradictory one—that I don't necessarily like water trading. I don't like the water trade because I don't like the fact it's being exploited. On the other hand, I make a very, very good living out of doing it. I know that sounds like a complete contradiction with everything I might say...Sadly, my job—if I don't do it, someone else will.

'Well, you can say, "Then step aside and let someone else do it"—if you have got the moral gumption and the position that you are stating. But unfortunately that is my family's livelihood. You have got to just keep moving and keeping up with the times. Sadly, some of us, in our journeys in life and making a livelihood, well that might not necessarily mean that you believe that everything about your business is totally ethical and sustainable, but you keep on doing it.'

Speaking in 2007, a trader shared Desiree's misgivings about the amount of money being made and the impact on rural communities, especially from the trade in permanent water.

'I'm in a difficult position,' the trader said. 'I'm against water trade and I do it. It's terrible, isn't it. Morally bankrupt. It's opportunistic.'

But one key trader we interviewed was entirely unapologetic about making money at farmers' expense.

'It's their own damn fault,' he said. 'It is "buyer and seller beware".'

The onus was on farmers, he claimed, to better understand the market, and to protect themselves from the worst types of gouging, such as by splitting their sales and purchases across different brokers and exchanges, and by portioning them out, so as not to send such a strong signal, or create such a lucrative opportunity for traders and brokers to exploit.

'I'm not entirely mercenary,' the trader said, 'and I'm not entirely in agreement with those guys that you're talking about, but there's an element of, you know, "You can't imagine it's still 1985 and think you're going to go places, guys. You've got to, you know, get with the times."'

CHAPTER 16

FRONT-MEN AND FRONT RUNNING

On any given day in the water market, farmers saw strange transactions and price movements. Flirting offers might flicker into life and then disappear just as quickly. Farmers also noticed quick changes in prices, usually upwards and for no apparent reason. And just as baffling were the static offers that stayed around, with fixed quantities, even through periods of drought—being drip-fed in what looked like a controlled way. The behaviour of prices and bids and offers was a mystery, but farmers suspected some kind of market manipulation was going on. They suspected the market was neither clean nor fair.

'If you go digging,' farmer Peter Burke said, 'you'll be horrified.'

The technological arms race already meant the market was arguably unfair. Water bots and supercomputers gave traders a permanent advantage. But the traders sought a further advantage—by bending and sometimes breaking the rules.

'The traders skate close to the edge,' broker Desiree Button said. 'And they skate over it.'

'If you need to just buy a little bit more to do your program,' farmer Laurie Beer said, 'well that's what you do. But I do have an issue with people running their business on trading water. Trying to trade water at absolute maximum profit and manipulating the market.'

In the water market underground there were whispers and allegations about sundry forms of misbehaviour: insider trading, misleading conduct, outright fraud. There were claims that traders could game the rules of individual exchanges, and use allocation trades and inter-valley trades to manipulate the market.

'Inter-valley trading is a lot of where the market failures are,' one insider told us, 'and it is where a lot of the rorting takes place.'

One particularly unsavoury tactic involved arbitrage, this time not with bots, but directly between brokers and farmers. Brokers would misrepresent the market price to clients, in order to solicit above-market bids and below-market asks, and then trade on them. This strategy was known as false arbitrage: for example, a broker might sell water to a client at $110 per megalitre when the true price was $90.

'As far as cowboy brokers go,' broker Clinton Davis said, 'there are plenty of them. Some of them, when dealing with farmers and novice investors, might not tell them all the rules and regulations. I shouldn't comment on that, but I see practices and I don't like them. They will say anything about where water can and can't go and what the returns might be. There's a lot of tricks. They will look and find where the market is. Say if it's $100 a megalitre, they'll come back and say, "I can find a buyer for you, but it's now only $80." They're just on about trying to manipulate.'

This tactic worked in both directions.

'The market is very volatile,' Desiree Button said, 'and you wonder why sometimes. Buying and selling and hedging—you can buy water and have it reselling at the other end. Or you can go in and say to someone, who probably wasn't quite up to speed—if they just relied on what you were saying—"Well, I can get you a thousand megs at a hundred bucks," with the knowledge that there is a whole lot of water there which is available at $90, you can scoop it at $90 and sell it for $100. Have a nice little $10,000 return in a single day on a single trade.'

In addition to building complex predictive models, and relying on computers and high-speed connections, the traders used a combination of legitimate and underhanded tactics to be first with price-sensitive information. In other words, rather than forecast or guess what farmers were planning to do, the traders could actually find things out—secret things—and then use that information against the farmers.

There was a 'hacking ethos' in many of the trading rooms. Using fair means and foul, the traders could gather price-sensitive data and metadata from insecure sources such as water exchanges, irrigation groups and the Bureau of Meteorology. (Despite the bureau's status as a government agency, and its important role in the water market, its website was not secure even as late as 2021. In 2015, foreign hackers infiltrated the site, obtaining access to valuable satellite imagery and weather data.)

Apart from gathering information in the digital world, the traders sought it in the physical world. They cosied up to leak-prone politicians and officials and market managers. They drove around, and they set up shop near exchanges and farming communities. They used drones and planes and Google Earth, and they hired trusted local front-men who could visit farms, join conversations in pubs, and otherwise tap the rural network and the water underground for valuable information.

When traders knew what farmers were about to do, they could 'front-run' the farmers' trading. This involved quickly buying water when they knew the farmers were about to buy, or quickly selling when they knew the farmers were about to sell. The intention was to ride the price up (after buying) or watch it fall (after selling) and then book a profit, and perhaps re-enter the market to buy or sell at an advantageous price. This was a type of 'information arbitrage', which involved trading between two states of the world: one in which the private information was secret, and one in which it was not.

The term 'front-running' comes from the old trading floors, such as at the New York Stock Exchange and the London Stock Exchange, where brokers or jobbers would literally run ahead of large pending orders so as to profit from their impact on prices. In the dark, over-the-counter derivatives markets, banks were widely suspected of using customers' information against them to front-run OTC trades. In the stock market, the 'flash boys' strategy of high-speed trading was a digital-era version of front-running. In that case, the time advantage from 'running' was measured in microseconds. Energy market players, too, had collected information from customers trading on platforms such as EnronOnline, and there was talk that they had been able to front-run customer trades. For water traders and brokers, the possibility of front-running was an enticing source of profit: a hidden margin on farmers' trades.

The scale of front-running in the water market is very difficult to prove and, so far as we are aware, there have been no cases or prosecutions so far. Worryingly, however, regulators and traders have confirmed that front-running in water is possible due to the lax controls on traders and brokers. And perhaps even more worrying:

by virtue of those weak controls, some varieties of front-running in the water market may not even be unlawful.

Real information was highly valuable for traders, but fake information was useful, too. Regulators and exchanges had limited control over the accuracy of trade price reporting. Through the tactical use of disclosures, traders could distort and manipulate market information to suit their own interests.

Traders could use fake trades, for example, to send a false price signal to the market. One way to do this was to report internal transactions (such as moving money between a trader's own accounts at a self-set price) as a 'market' trade. Traders could also place buy and sell orders and then withdraw them before they were executed. (Feigning interest in or demand for a parcel of water is called 'spoofing'.) Such false signals were useful ways to lower or raise price expectations as traders needed.

'Sometimes,' Ben Dal Broi said, 'you get that sickening feeling that certain traders are playing us. These ones will advertise water at an attractive price. When you call, they'll say, "Sorry mate, you just missed out. But, if you're quick I've got another parcel here for only $20 a megalitre more..."'

Another way to game markets and reporting systems was to strategically misreport trades. As the ACCC found in 2020, brokers could influence the market price 'by misrepresenting the price on trade approval applications to signal to the market (via the state register) that the price of a water right is higher than is actually the case'.

'Information in the water registers is not audited,' an exchange manager told us. 'Traders can report junk data and there are few if any consequences.'

Yet another way to send false signals into the market was to simply not report trades at all. Many large trades—which are themselves market-sensitive information—were not reported, even though their existence and details were of course known to some people, including the parties to the trades. Under-reporting seemed to be widespread, as was the reporting of zero-dollar trades, which represented a variety of different instances of strategic misreporting and non-compliance.

False prices were not the only strategic falsehood in the water market. We suspected pseudonyms were used to conceal the identities of investors and traders, and the details of specific accounts and transactions. In the water underground, banks and agribusinesses were rumoured to have borrowed names from popular culture—Ferris Bueller, Dirk Diggler, Eddie McGuire, the Leyland Brothers—and from history and geography—Bonaparte, Xerxes, Delhi, Billabong.

In addition to fake prices and identities, the traders' tactical toolkit included strategic disclosures about trading intentions and market conditions. The daily reality (or unreality) of the water market was a jarring experience for farmers. Public information releases and announcements often did not accord with farmers' direct experiences of rainfall and water availability, for example. Farmers suspected disclosures were used strategically to influence perceptions—of plans, flows and values—and therefore water prices.

On 16 October 2020, farmer Laurie Beer emailed us about a strange sequence of announcements.

> You may be interested to see this release from NSW Water on the 1st of the month, which many of us took to mean further increases in allocation would be unlikely until there was further significant rain. Then, just two weeks later, they announced a 9 per cent increase in our allocation when there had been no

> significant inflows. In fact the storages had been rising consistently at a steady rate for the whole of winter. Could this be someone withholding or providing misleading information?

Some disclosures from brokers and agribusinesses and banks seemed to be strategic. It is possible that some official releases were strategic, too, but more likely they were stale and clumsy, not deliberately misleading. It was common for public officials to misunderstand how their choices of wording and the details of their disclosures would play out once they reached the market. The officials and traders lived in strikingly different worlds—and this was one more source of advantage for the traders.

Another striking feature of the water market: with all the money being made, and all the ambiguity and opacity about what was going on, traders routinely lied about what they were doing and why. How they traded, how often they traded, who they traded for, and the impact they were having on the market. Falsity became as much a routine feature of the market as the mind-straining complexity, the lack of transparency, the arbitrage and the bots.

CHAPTER 17
UNPROTECTED

From our work in finance, including our research on the Hayne royal commission, we knew that powerful protections were in place for participants in Australia's financial markets. The *Corporations Act 2001* is the principal piece of Australian legislation for ensuring the integrity of companies and markets. It aims to keep everything above board, and over the years it has sent many people to jail.

Chapter 7 of the act regulates the financial services industry. A holder of an Australian financial services licence must, among other things, do all that is necessary to ensure the relevant financial services are provided efficiently, honestly and fairly; manage conflicts of interest, including by disclosing, controlling and avoiding them; protect clients' money and other property; and keep adequate records.

Division 2 of part 7.10 of this chapter of the act deals with market misconduct and other prohibited actions relating to financial products and financial services. Most of these actions attract significant criminal sanctions and civil penalties. The following types of misconduct are specifically banned: market manipulation; false trading and 'market rigging'—such as by creating a false or misleading appearance of active trading, or by artificially maintaining a trading price; making false or misleading statements; inducing persons to deal; dishonest conduct; and misleading or deceptive conduct.

ASIC is the independent supervisor of Australia's financial markets. It wields an arsenal of Corporations Act powers, including the power to conduct surveillance and otherwise to monitor holders of financial services licences. ASIC aims to promote 'fair and efficient' markets, which ASIC interprets as ones that feature integrity and transparency, and in which investors and consumers can participate in a confident and informed way. Under the Corporations Act, ASIC has issued specific integrity rules for securities markets, including the ASX. The ASIC Market Integrity Rules (Securities Markets) 2017 set out obligations and prohibitions for conduct on these markets.

The Securities Market Rules establish pre-trade and post-trade transparency obligations; procedures for managing conflicts of interest; a 'best execution' obligation (which means, broadly speaking, that a broker or trader acting as agent for another must execute the trade in the best available way); penalties for the false or misleading appearance of active trading; requirements for fairness and transparency in allocating orders; and obligations to keep records of dealings on the licensee's own account.

Financial market offences attract criminal sanctions. And following the Hayne royal commission, the *Treasury Laws Amendment (Strengthening Corporate and Financial Sector Penalties) Act 2019* made the penalties for financial market misconduct even stronger.

To examine why these protections were not enough to prevent

misconduct in the water market, we reached out to legal friends: barristers and solicitors, several of whom had previously worked in or for government and in the water sector, and had seen every type of scandal and every kind of defendant. And we turned again to our underground water network.

Why was there so much fraud and misconduct in the water market?

'You're not going to believe my answer,' one lawyer said.

In the design of the market, she told us, there had been little consideration of what participants would do once inside. Hardly any of the National Water Initiative or the Murray–Darling Basin Plan, for example, was concerned with matters of market conduct. But one decision of the market designers did concern conduct in a very direct sense.

Through a deliberate act of government, the water market wasn't regulated the same way as other financial markets. The participants in the water market were explicitly excluded from the normal controls and protections that were available elsewhere. Specifically, the *Corporations Amendment (Water Trading Exemptions) Regulation 2014* excluded tradable water rights, and arrangements to buy and sell them, from the definitions of 'derivative' and 'financial product' under the ASIC Act and the Corporations Act.

In the explanatory memorandum for the 2014 amendment, the Australian Treasury outlined the consequences of that decision:

> the Australian Securities and Investments Commission (ASIC) will not regulate such arrangements as derivatives for the purposes of the Chapter 7 of the Act. The Regulation will also ensure that ASIC will not be responsible for regulating them as financial products under the ASIC Act.

> Consequentially, certain water market intermediaries facilitating trade in basic tradeable water rights will not be required to hold an Australian Financial Services Licence (AFSL) and operators of markets in basic tradeable water rights will not be required to hold an Australian Market Licence (AML) and will not be subject to regulation by ASIC as a financial market.

The amendment had other practical implications, including for how water trades were 'cleared' and settled. Excluding tradable water rights from the definition of derivatives meant they did not require a centralised facility or mechanism for clearing and settling market transactions. Water exchanges were not required to hold an Australian Clearing and Settlement Facility Licence, and were not subject to regulation by ASIC as a clearing or settlement facility.

There was another important legal distinction, too. In most Basin jurisdictions, farmers do not own water resources as legal property. This affects how the water market is regulated, and it affects the financing of family farms. As Waterfind director Tom Rooney told the *New Daily* in March 2019, farmers can't easily borrow against their water rights: 'Back in 2004, the government did a half-arsed job separating water and land. As a result, water rights aren't bankable... The only way they can get money from them is to sell them.' He also thought water rights needed 'to be recognised as genuine, definable assets for farmers to more easily borrow against them'.

Victoria's 2004 *Securing Our Water Future Together* policy paper had claimed unbundling of water rights would 'reduce borrowing costs, by providing for mortgages directly over water'. But if banks would not accept water rights as collateral, farmers could not borrow against them. The result was a systemic bias against family farms.

And just as farmers couldn't use their water as collateral, there were barriers, too, to them using water as superannuation.

'You are not allowed to do that,' Ben Dal Broi said. 'When I asked my accountants, they nearly lost their biscuits. "No, no," they said, "using your super to buy water for use in your own irrigation business is against the law." But people are doing it, even though you are not meant to, and some brokers are willing to make it happen. One broker said to me, "Sure, we can make it happen, we sell the water you hold in your super account to someone else and give you a sweetheart deal selling you back someone else's water." I can buy a farm with my super if I wanted to, so why not water? It puts me at a financial disadvantage.'

The Australian water market had become a financial market, but it was not regulated as one. And nor did the Basin water exchanges apply trading rules comparable to those in other markets. Tactics strictly prohibited elsewhere—such as in the markets for stocks, bonds, commodities, electricity and real estate—were permitted in the water market.

The water market rules did not, for example, prohibit market manipulation—an extraordinary fact. With the crucial protections established under the Corporations Act and the ASIC Act not available to farmers and other participants, the water market was primed for tactics that, if perpetrated elsewhere, would send the perpetrators to prison.

'It almost feels like "anything goes" in this market,' a Basin businessman told us.

The water market's main features—prices, information, participants—were transparent neither to outsiders nor to most insiders. Politicians and policymakers had turned irrigation water into a tradable, financialised product, and the market had become ripe for exploitation. Seeing how accommodating the water market controls were, the international hedge funds and investment banks couldn't believe their luck. They'd readily transported to Australia's water sector the aggressive strategies they were using in global financial and commodity markets, including in dark market swaps and other derivatives. And in Australian water, they could go even further.

CHAPTER 18

WHO DO YOU WORK FOR?

On the water market, brokers perform an analogous role to those on the stock market, but there are important differences. Compared with stockbrokers, there is little scrutiny or monitoring of the conduct of water brokers, whose activities are ill-defined and under-regulated. For instance, the line between water brokers and water traders can be fuzzy. The primary difference is that brokers usually act for someone else, whereas traders have their own capital or are directly employed to trade the capital of a bank, a fund or another cashed-up participant. But in practice, brokers sometimes trade on their own account; sometimes they are tied to banks and funds; and sometimes they play both sides of a transaction.

There are other differences, too. The price of entry into water broking is very low. We heard from the underground water network that water brokers had to fulfil no qualification requirements. There had been few professional bodies, at least initially, and professional standards and networks were much less developed than in other areas of finance. Crucially, there was no binding, industry-wide code of conduct, nor were there common integrity standards or commitments.

In the 2010s, journalists Adele Ferguson and Alan Kohler had looked at Australia's system for retirement savings and financial advice. They discovered a quagmire of conflicting interests. The most damaging conflict of them all: financial advisers were ostensibly working for clients—but all the while they were receiving commissions from the banks and funds. In Kohler's words, retail customers were being 'harvested by financial advisers and the wealth management industry as a matter of routine, like so many ripe ears of corn'.

This conflict of interest became a focus of the 2019 Royal Commission into Misconduct in the Banking, Superannuation and Financial Services Industry. As Commissioner Kenneth Hayne AC QC pointed out in his interim report, '[M]ost financial advisers came from a background of life insurance, in which a sales-driven, commission-based culture prevailed...These being the roots of today's financial planning industry, the culture has endured.'

The industry was meant to support people making life-changing decisions, but it was actually configured to sell them investment products, on behalf of and for the benefit of financial institutions that were in the background. AMP and the banks referred to this model as 'advice-based distribution'. The income of planners depended on how much they sold.

The advice industry hid behind the language of 'help' and 'advice', and the false perception of their salespeople as 'advisers' and 'planners'. As Kohler wrote:

> the planner always spins an excellent web of fancy talk about 'risk tolerance' and what's happening with the markets, and some of the important principles of investing. But it's really like talking to a TV salesperson at JB Hi-Fi or a car salesperson at the local Toyota dealer: they know what they're talking about, and sometimes they're brilliant at it, but their job is to sell you something.

This crucial system was broken. The royal commission looked closely at the deep conflict of interest, which had reduced confidence in financial services and the integrity of Australia's system of retirement savings.

'Experience shows that conflicts between duty and interest can seldom be managed; self-interest will almost always trump duty,' wrote Commissioner Hayne in his final report. He made wide-ranging recommendations to lift the integrity of financial planning, broking and advice, including by requiring financial advisers to disclose their lack of impartiality, and ensuring mortgage brokers acted in borrowers' best interests.

In looking closely at water brokers, we found very much the same issues that had been front and centre in the Hayne royal commission. Like financial advisers and mortgage brokers, water brokers had a conflict of interest. The old problem—of people who looked like advisers but were actually salespeople—was prominent in the water market.

We heard from our water network that there were barely any restrictions on what brokers could do and who they could work for. Water brokers were also not subject to tight rules about insurance or safely managing clients' funds, such as via statutory trust accounts, fidelity or assurance funds.

As a result, conflicting interests were rampant.

There were other parallels with the royal commission. Just as there is a gaggle of regulators in water, Commissioner Hayne found the same thing in finance. ASIC, the ACCC and the Australian Prudential Regulation Authority (APRA), for example, had overlapping roles and functions, and tripped over each other in their efforts to detect and pursue financial misconduct and market failures.

Another parallel was the complexity of the market, and the opacity of market information. Commissioner Hayne was highly critical of the assumption that non-specialists could read and absorb pages and pages of financial fine print that seemed almost as though it had been written with an intent to confuse. In the water market, too, non-specialists were drowning in the fine print. But the conflict of interest was the main reason farmers were wary of using brokers.

'Farmers use more brokers than they trust,' farmer Peter Burke said.

Some brokers—probably the majority—were honest and well meaning. But the temptations were large, and we heard it was not uncommon for brokers to make the most of the lax regulations. The less scrupulous brokers partnered up with traders to use information in ways that didn't benefit the farmers. There were few if any consequences for brokers who flouted the rules. And the least scrupulous ones, it seemed, tended to make the most money.

In the stock market, there were strict rules about insider trading and other aspects of the misuse of information. In the financialised water market, however, there were few controls to prevent insider trading, few controls over brokers' use of information, and scant rules about how and by whom brokers would be paid. Roles were fuzzy, monitoring was weak, and there was little to stop brokers from sharing information with traders so the traders could profit from it.

Operating through exchanges and also directly (that is, 'off-market'), brokers could use customers' information against them—such as to front-run customer trades or take counter-positions against them in the market. (Brokers were known to have traded on their own account against clients whose interests were supposed to come first.) Or they could simply sell customers' information for secret payments and commissions from big traders and investors.

'Right from the beginning,' one official said, 'regulators downplayed the danger of insider trading.'

'The regulators and the rules are useless,' a lawyer told us.

There have been complaints of insider trading in the water market, and even prosecutions of water brokers, but the detected offences have mainly been generic, such as those for fraud or theft of client funds. Without specific legislation, and without the financial market protections of the Corporations Act, taking legal action is difficult. Relying on common law or consumer protection law to pursue unconscionable or misleading conduct in the water market is also difficult.

'It's a bloody mess,' another lawyer said.

By this stage in our journey, we'd learnt a lot, and we felt we had a good sense of what was happening in the water market, and what precisely had gone wrong. The decisions to allow anyone in, and then to remove legal protections for participants, had created a free-for-all that was not tempered by the role of water brokers.

But, as farmer Lachlan Marshall had warned us, our journey was only just beginning. There would be more doors to go through and more shocks to come. In the water market underground, we heard intriguing whispers of a bigger strategy and an uglier truth, made possible by the poor market design and lax oversight.

As we delved further, the water market insiders shared with us something normally only uttered at the back bars of the Globe and Exchange hotels in Deniliquin, or around the snooker tables at the Termo in Tocumwal. The newcomers to the Basin had worked out a way to supercharge their trading strategies into a customised money-machine that would take their profits to a whole new level. And in doing so, they would change the shape of Basin agriculture and Basin communities.

PART IV: CORNERING THE MARKET

CHAPTER 19
A NIGHTMARE SCENARIO

From the beginning of widespread water trading in the Basin, there were whispers in the water market underground about a larger strategy. Farmers feared banks, big investors, agribusinesses and other corporate interests were buying up water in a big way and might be able to corner the water market.

The idea of cornering a market refers to having a significant degree of control over supply, demand or prices. Equivalent terminology includes being able to exert market power or monopoly power. Cornering, in short, meant conquering.

One farmer, speaking in 2007, expressed his fear this way: 'I was talking to some fairly influential people who had done research on where water will be in 2010. With unbundling, this person said, he can foresee that people like superannuation funds and insurance companies will buy large tracts of water, then sit on it, and then sell it to towns and will make very good returns on it.'

Through our research, we discovered these concerns were well founded. The fear—expressed by our contacts and interviewees with an exceptional degree of consistency—lines up closely with the opportunities and the tactics of the banks and quasi-banks in the water market. It also fits the data on water prices and the predominant flow of water rights.

Just as important, it lines up well with experiences in other markets. For banks and hedge funds and other major traders, cornering held the promise of supernormal profits, and the traders could attempt it with an amount of capital that, for them, was modest in scale. They had cornered other financial markets in broadly the same way. In other words, the big traders had motive and opportunity, and they had form.

Once we understood how the cornering strategy worked, we finally discovered what had happened in Australian water. At the heart of the policy experiment, at the heart of the regulators' analysis, at the heart of scores of official reports and plans and strategies, there was a simple, basic error.

How to corner a market, step one: organise.

Not long before the collapse of Bear Stearns and Lehman Brothers, the large water traders realised they could profit by aligning their strategies and presenting a common front. In their engagement with Basin governments and regulators, the large traders presented a shared position on technical matters and significant market reforms. The traders pushed jointly for the removal of geographical restrictions on the movement of water rights, the removal of limits on non-irrigator ownership of those rights, and the lifting of any other restrictions on outside participation in the market.

They also advocated for the status quo in which normal financial market protections would not apply to the water market, and in which water traders and brokers would not require a financial services licence.

There were other ways, too, in which the traders could collaborate. Lax regulatory oversight in the market created the potential for coordination and even collusion between big players in their day-to-day trading. There was an economic basis for this mode of coordination. Collectively, the big traders had shared interests and they adopted the same or similar methods. And there was a practical basis, too.

The major traders had already formed bilateral and multilateral relationships with each other, including by collaborating in other markets in other countries. The banks and hedge funds worked together in industry groups and forums, for example, and in back-office systems and institutions such as those for foreign exchange settlement. They struck deals with each other in the murky, invitation-only trading environments known as 'dark pools'.

The banks and hedge funds were also members of special clubs, such as those that allowed them to trade with one another. The International Swaps and Derivatives Association, or ISDA, is an example of such a club. Founded in 1985, it was made famous in *The Big Short*. Fans of the book and film know that elusive and expensive ISDA accreditation was the hunting licence sought by protagonists Charlie and Jamie so they could bet against the US housing market with credit default options.

Relationships such as these provided a mechanism for coordination here, too, as did direct and indirect market signals, and formal and informal overtures, just like the ones used by colluding participants in the first telecommunications spectrum auctions. The banks themselves also had an international track record of collusion. They'd been caught in cooperative schemes to manipulate the foreign exchange market, the short-term money market, commodities markets and other markets. The water market was ripe for the same mode of collusion.

The word 'monopoly' has often been used to describe how the major traders have been able to operate in the water market. As farmer Peter Burke put it: 'When they let all these investors in, the whole footprint changed. A small number of powerful players are able to act like a monopoly.' We also heard this from a Basin water trader: 'Traders do have a degree of monopoly power, partly through hiding information, but partly through using their buying power.'

In 2003, ABC's *Four Corners* program broadcast its Walkley Award–winning investigation 'Sold Down the River', presented by Ticky Fullerton. 'Picture an entrepreneur,' Fullerton said, 'plying his business from a mid-city cafe...What this silvertail buys and sells is a low-maintenance but now hot commodity: the right to use megalitres and gigalitres of water from a meandering brown river thousands of kilometres away. Drought boosts his price; rain depresses it. Not so long ago, the water was free for all to use. Welcome to the world of the water baron.'

The Victorian Farmers Federation also used this term in a submission to the ACCC in 2019: 'The VFF never supported the removal of the 10 per cent non-landholder limit. The VFF believe it provided sufficient protection from water barons manipulating the water market.'

In industrial economics and game theory, an oligopoly that acts like a monopoly is called a 'collusive oligopoly'. Another word for it is 'cartel'. Is there or could there be a cartel in Australian water?

In economic policy and regulatory circles, coordination between major market players is notorious for being difficult to detect. The signals and overtures they convey to each other can be subtle, and collusive and monopolistic behaviours are hard to distinguish from mirroring and mimicking ones. The question always has to be asked: are the big players doing the same or complementary things because they are deliberately coordinating, or are they innocently copying each other? Or have they come up with the same strategies independently?

Another problem in detecting monopolistic conduct in water: the major Australian water regulators and government departments have consistently and actively downplayed the possibility, just as they downplayed the likelihood of insider trading. On 26 May 2017, David Parker, deputy secretary of the federal Department of Agriculture and Water Resources, told the Rural and Regional Affairs and Transport Legislation Committee:

> since land and water were separated, there has always been an element of the community that was against that…and some fears in the community that you would see water barons and so forth who owned all the water emerge and that, the farmers having sold their water… they would then be subject to exploitation by these so-called water barons. We have not actually seen that emerge. One of the reasons for that is, if you hold a water entitlement, and you do not hold land to use it on, and you get an allocation against that entitlement, it is of no use to you unless you sell it. What do you do? You sell it back into agriculture purposes.

But a collusive oligopoly was certainly feasible, due to weak restrictions and scant oversight of market manipulation and collusive conduct. That fact was good news for the larger traders, who aspired to control markets in general, and for whom the idea of cornering the water market in particular was a bewitching goal. But at this early stage, no amount of coordination could have turned the big traders into an effective cartel, or any kind of oligopoly. They needed more arrows in their quiver. They needed to simplify the task, and that is precisely what they were able to do. The big traders could step towards their goal of controlling the water market in a way that the market designers never expected.

CHAPTER 20
PLAY MONEY

A striking feature of Wall Street is how all the different types of financial institutions are sorted into a rigid hierarchy. The bulge-bracket investment banks are at the top of the pecking order, then hedge funds, followed by commercial banks, and then fund managers. Mortgage lenders such as S&Ls are a long, long way down.

Within each of the categories there are further sub-hierarchies. Some investment banks are more powerful and prestigious than others, while the broad category of 'fund managers' encompasses a full spectrum of types, from glamorous, high-wealth funds at the top, to workaday, low-service retail funds at the bottom.

There are even strict hierarchies within the banks themselves: in investment banks, for example, dealers look down on the pointy heads of corporate finance, who in turn disdain the research analysts writing broking reports.

The hierarchy is important in several respects. It affects where people work: the smartest and most ambitious dealers compete for jobs at the most prestigious houses. The hierarchy also affects the conduct of individual staff: their arrogance and swagger, what they wear, where they eat, and—most importantly for them—what they earn. And it affects how deals are carved up, and which parts of the financial system are permitted to deal in which types of assets and to pursue which levels of returns.

The banks and quasi-banks that transplanted Wall Street methods into the Australian water market also imported the strict Wall Street hierarchy. And this meant carving up the different types of opportunity and the different scales of financial returns.

In the financial hierarchy, investment banks and hedge funds (along with so-called 'high-wealth individuals'—i.e. very rich people) walk right past investments that offer anything like the ordinary rate of return. Normally, for example, these banks and funds would pay no attention to big, long-term agricultural investments that offer returns of 5 or 10 or 12 per cent. Instead, the yield-hungry banks and funds chase returns of 20 to 30 per cent or more.

Paradoxically, however, in the Australian water market, the top-tier banks and hedge funds teamed up with pedestrian pension funds and super funds and university endowments and 'managed investment schemes'. These relationships were unexpectedly important. For the big water traders, they were the key to a vast treasure.

Playing the market on a much larger scale would require financial gymnastics on the part of the banks and quasi-banks who'd learnt the ropes on Wall Street. Using a variety of structures, these big traders partnered with agribusinesses and fund managers—and then they went to the financial markets to solicit lazy, low-return capital from wholesale and retail investors.

Pension funds from China and Canada, and superannuation funds from Australia, were happy to pursue investments that offered a low-risk, 5 to 10 per cent annual return, over a horizon of ten years or so. They stepped forward in a big way to invest in Australian water. (In practice, the returns were often even lower than 5 per cent. In 2016, RMIT senior lecturer in management Dr Charlie Huang, who researches Asian investment in Australia, found that external investors had received poor returns on their agribusiness ventures. 'I interviewed Chinese investors with long-term investments in Australia,' he told banking newsletter *BlueNotes*, 'and when I asked them about the return on investment, to my surprise it was very, very low.')

'VicSuper is in the market,' a broker told us. 'It owns hundreds of thousands of megs as well as a lot of land through its Kilter fund.'

In 2006, VicSuper, in partnership with Kilter Rural, launched the Future Farming Landscapes project, which used $200 million to purchase 9,000 hectares of land and 85 gigalitres of permanent water. At the time of writing, the fund had 4,500 hectares of redeveloped and irrigated farming land plus 4,300 hectares protected for ecosystem services. These assets were worth $430 million in total.

VicSuper Ecosystem Services is Victoria's second-highest ranked water holder, with 56.5 gigalitres of high reliability water entitlements. The biggest owner is another super fund, Canada's Public Sector Pension Investment Board, with 78.1 gigalitres. According to the *Weekly Times*, that parcel 'is worth about $508 million and is held by a $1 company the Canadians set up in October [2019] called Fresh Country Farms of Australia No. 4'.

Most of the lazy money was channelled into new industrial-scale farms where hydroponic wizardry turned dustbowls into investment brochure–friendly fields of olive trees and almond trees.

Between 2006 and 2018, Australian orchard plantings doubled in size, from around 24,000 hectares to around 48,000 hectares. And between 1997 and 2018, total irrigation development in the Mallee Catchment (i.e. below the Barmah Choke) also doubled, from around 40,000 hectares to more than 81,000 hectares.

For the pension funds and other conventional investors, the case for this type of investment was compelling. Australia's water market was the poster child for how market mechanisms could be used to solve public policy problems. The market was said to be driving efficient investments in formerly arid regions, where the horizons were wide, the scale was large and the land was cheap. It was much easier and more cost-effective to buy land there and to plant new trees, rather than buy existing farms upriver.

For the foreign investors, there were other sweeteners, too. The olives and almonds were a kibbutz-style effort to green the desert and feed the world. The investors could tell a story about 'ethical investment' and, better still, they could reap tax advantages by operating across national borders.

For the banks and hedge funds, too, these arid investments had fertile attractions. The partnerships they entered into with wholesale and retail fund managers were odd hybrids of fast and slow money. In some cases *de jure*, and in all cases *de facto*, the partnerships were portfolios: multi-divisional enterprises, consisting of long-horizon agricultural investments and short-horizon water trading.

'Why would bankers care about marginal, low-return investments?' one financial market insider asked. 'Vast, arid fields of almonds and olives are just not sexy enough. Not worth getting out of bed for.'

But the megafarms, established predominantly below the Choke, owned huge quantities of water rights.

'They are planting all these orchards,' another insider said, 'but their real interest is in the water and how they can trade it.'

Having partnered with the big traders, the megafarms held their water in the optimal portfolio for maximising the impact on the market: a combination of permanent water and up to one-third temporary water, much of which was available for active trading.

'The megafarms and megatraders arrived at around the same time,' the insider said. 'This wasn't a coincidence. It was a strategy.'

'There are 7,500 hectares of almonds,' a Basin farmer said in the 2007 AgriFutures report on water trading, 'and 900 hectares of olives, and half of them would be less than three years old. It's just amazing. So there is a lot of water that is actually allocated that isn't actually being used there yet. My understanding is that a few years ago there were some people—there was a bit of leasing of water going on while there have been occasions to establish development. They're changed now. I think now they are in the situation to purchase the water, buy it all up front, and then put it back on the temporary transfer market.

'For the last three years, I've had an arrangement with some of the almond growers in Melbourne. It's like a contract: we agreed on a price and that price stood for three years, which over that time has actually doubled what the water was going for, so we actually have done all right out of it. Now the almond industry has switched around; now they're actually buying it all up front at a lower price and trading it back themselves now. I'm not actually sure what they are doing. Our agreement ran out last year and so I haven't heard from them.'

The megafarms also came with the capacity to store water. 'A lot of these large developments,' another farmer said, 'are putting in 500-megalitre dams as buffer storage, but small growers can't do that.'

The megafarms' legitimate, everyday purchases of water were extremely useful for the water traders, who were highly attuned to volatility and to the market impact of information and large transactions. The legitimate purchases affected water prices, and the trader-partners could incorporate those price movements into their own information-based trading strategies and programs.

In this way, the traders could influence the water market with water that wasn't on their own balance sheet. This strategy wasn't especially efficient in water, but it was highly efficient in capital.

Even more importantly—and lucratively—the big, lazy-money investments achieved a radical heightening of water scarcity. They created large blocks of water demand—and these, too, could be controlled and used tactically by traders in the market. For the banks and hedge funds, rather than investing in speed or information, the diversified partnerships were an investment in market power. Thus able to influence prices and exert market clout, the traders could make a killing.

'It all gets back to the yield on our water allocation,' farmer Laurie Beer said. 'Back in the 1980s and '90s, we would've considered an 80 per cent average yield as pretty conservative. I mean you didn't get 80 per cent every year. Some years, you got over 100 per cent and some years you might only get 50. On average, over ten years, you'd say you probably got an 80 per cent yield on the amount of water that you actually owned.

'Well now, the latest figures, in the last ten years, I think you'd be flat out even calling it 50 per cent yield on those same entitlements. And back when I was still going to high school, I can remember the starting allocation one year was only going to be 50 per cent. And everyone thought that was an absolute disaster. No one had heard of that so low, to start our allocation.

'Well, now our norm is to start at zero per cent. And not even get above 50 per cent. It's happening quite regularly. Well, it just didn't happen 20 or 30 years ago. And I mean, I don't believe the yield of our catchment has changed very much. It's where the water's going. It's a challenge, where it's being allocated.

'I feel that there's some big players in the market that are able to manipulate the water market, and I feel they are doing that, especially in scarce years. When there's plenty of water about they won't be able to, but when it's scarce I think some of the big players are able to actually manipulate the market to their advantage. There's some big players you hear about, but the water market's not transparent enough to know for sure who they are. They can do it through a third party or a Caymans holding company or something like that. And you don't really know who it is.'

Other farmers, too, said the big blocks of fixed demand for water downstream were driving the market dynamics, reducing security and increasing pressure on the limited channel capacity of the Murray.

Crucially for the banks and hedge funds, the mass-scale investments in novel crops directly complemented their active day-to-day trading. The big traders owned large amounts of water directly, and they could control a much larger amount via the megafarm partnerships. Scarcity was magnified and so was market power.

'This is not how the system was supposed to work,' water broker and consultant Gary Weber told us. 'Farming family irrigators are being killed.'

CHAPTER 21
PINCER MOVEMENT

In most financial markets, there is no fixed limit on the number of securities that can be traded. Company shares, for example, and over-the-counter (OTC) swaps and derivatives are not fixed in their supply. But Basin irrigation water is traded under a strict cap, and the way this is managed has affected the viability of the megafarm strategy. Specifically, the way the cap and environmental water were historically defined allowed the traders to magnify the impact of scarcity in the market, and to pursue profits in a particularly powerful way.

Rather than define environmental water in relation to environmental outcomes—such as stream health, tree health or biodiversity—the water market designers initially set the environmental allocation as a specific amount of water within a hard cap. Publicly announced, the cap placed a known limit on entitlements for irrigation and speculation. This left the market vulnerable to a giant pincer movement: creating false scarcity from below (using the big blocks of demand) pushing against the hard boundary of the cap above. The cap made the job of cornering the market much easier: there was a chance the whole market could be squeezed in a giant vice.

'The Murray–Darling Basin Plan is one component,' water broker and consultant Gary Weber said. 'The Plan tends to get conflated with a whole bunch of other reforms that have happened. The Plan itself is really only about the acquisition of water for the environment. But it tends to get discussed as though the overall reform picture is the Plan. I think if you went and started again, you might have targets that are not volumetrically based but more outcome-based. But it wasn't a scientific policy approach. It was a political compromise on numbers.

'The greenies want four-and-a-half or ten or whatever the hell they wanted. The farmers want zero. Let's pick a point in the middle! Which is just the art of politics, isn't it. It was a political compromise and doing it volumetrically is not the way that it should have been done. But it is the way it was done.'

The practice of buying back water for the environment also affected scarcity, and the viability of the pincer strategy, as did the unregulated capture of water in the northern Basin, such as through floodplain harvesting.

'The Murray–Darling Basin Authority spent $13 billion of taxpayers' money to buy water,' farmer Chris Brooks said. 'It was a real swiss cheese approach to buy swathes of water everywhere—and they've effectively gifted it to those knobheads in the north by letting them take an extra 3 million megs out without regulation. They just take whatever volume they want—unmeasured, unmetered, unmonitored, unregulated—so much so that there is now not enough water going down the Darling River. That's the stupidity of it. And it's effectively ruined the environment in the Darling. It's wrong for that environment, the Menindee Lakes. And they are not making their contribution to South Australia, which then forces pressure on the Murray. It's a debacle.'

'People made vast sums of money,' farmer Robyn Wheeler said, 'buying and selling water, and vast sums of money when these buybacks were on. And talk about insider trading! I mean, my God, so I find all of it offensive. Wrong. And then there's also the issue of hiding the water and the amount that has been paid. The amount through buybacks alone that has been paid to pretty big private companies. It's really, really concerning. You should be looking at it, a bit like the block of land near Sydney airport.

'The problem for Deniliquin is, under the ridiculous water sharing arrangements, when there is no water in the Darling, we don't get any water. And there is no water in the Darling ever anymore, because it gets sucked out every time there is a flood. And it's all being drained to go down to South Australia. The Darling River system in its heyday had steamboats on it. It was just an extraordinary river system. And now it never gets a chance to recover, because whenever there are flood events, the water just gets taken out by very big commercial interests up north. It's being sucked out at two ends.'

Climate change is an even more powerful driver of scarcity—and this, too, is complementing the pincer.

In 2014, the UN Intergovernmental Panel on Climate Change (IPCC) released its fifth assessment report, which confirmed warming will continue through the 21st century. Based on projections for warming at 2°C, south-eastern Australia may experience a 40 per cent decline in annual water runoff. The IPCC found that, under a medium emissions scenario, the Murray–Darling Basin is likely to experience water security problems by 2050. These findings were confirmed in the 2018 IPCC report.

Recent research led by Professor Will Steffen, and published as the Climate Council report *Deluge and Drought: Australia's Water Security in a Changing Climate*, examined how climate change was affecting rainfall in the Murray–Darling system. Professor Steffen and his colleagues found that hotter conditions and reduced rainfall over recent decades had led to less runoff into streams, rivers, lakes and dams. South-east Australia had experienced a 15 per cent decline in late autumn and early winter rainfall, and a 25 per cent decline in average rainfall in April and May.

Though overall Basin inflows have fallen since the mid-1990s, water yields have fallen even more. This difference reflects impacts from moving water across the system (such as transfer losses), differential treatment of water in different regions (such as unregulated capture in the north) and the impact of market anomalies and unintended consequences (such as from waking up 'sleeper' and 'dozer' licences). Traders and their modellers know the likely impact of climate change on rain and inflows, and therefore on water scarcity and the dynamics of the water market. This is valuable input when deciding trading tactics.

The designers of the water market had thought a cap would save the Basin. Instead, aided and abetted by climate change and increasing water scarcity, the cap gave the traders a hard edge against which to conquer the market.

Just like the big energy traders had done in the 1990s and 2000s, the big water traders cheered when shortages were sharpest. (The energy traders were found to have used all sorts of underhanded tactics to magnify scarcity, such as making secret deals with power producers to order power-plant shutdowns at useful times.) Traders valued scarcity in and of itself, even more so than the water. Scarcity was a route to profit, whereas abundant water was a disappointment.

This is the essence of the difference between farmers and traders. Farmers made money—to the extent they could—by investing and taking risks and creating wealth. The traders got rich by skimming and gouging wealth from the farmers.

The megafarm strategy had a third dividend. Apart from the huge holdings of water, and the impact on scarcity, the industrial-scale farms gave their high-powered partners a cover story. A fig leaf. A licence to play. When the traders used the blended strategy—combining slow-money megafarms and fast-money water trading—they could argue their market transactions were connected to their parallel farming investments. A major trader explained the value of the megafarm strategy for traders who wanted to conceal what they were doing in the market.

'We can always wrap a bullshit story around our trading,' he said. 'I've handled pretty significant amounts of cash, and could always hide it behind an agricultural use. You know, I could concoct a story to support it. "Hey, we needed to buy that 10 gigs—and then the board made a decision, so we went and sold it back," or whatever. I could tell some fibs, if that makes sense, so I'm not simply trading, I've got a legitimate excuse to justify owning that water and acting in the market.'

The fast–slow strategy was the traders' Holy Grail: it was the focus of their quest, and it gave them the key to eternal life. And the perpetrators of the strategy—who had created the new water reality of lords, barons and serfs—were not unlike the Templars, because a lot of what they did was shrouded in secrecy.

CHAPTER 22
WALK AWAY

The big water traders had taken two key steps towards cornering the market. They'd coordinated their strategies, and they'd magnified scarcity, such as by owning a lot of water and by partnering with lazy-money funds and megafarm agribusinesses. Those partnerships came with play money, and play water. The design of the market, including how the overall cap and environmental allocations had been defined, made the task of amplifying scarcity easier, and it offered the possibility of grasping the whole market in a giant pincer movement.

The next crucial step in cornering the water market was to mimic a strategy that had yielded enormous profits for banks in the OTC derivatives market, and for traders in the energy market.

Derivatives were mostly traded in public exchanges—until the 1990s, when Wall Street banks realised they could control the derivatives market the same way they controlled the securitised mortgages market. This realisation was the genesis of the OTC market, which featured customised products and bilateral trading that was not subject to scrutiny the way public exchanges were. In OTC dark markets, the banks could set margins, charge fees, and 'price discriminate'.

Price discrimination is an important concept in economics and finance. It refers to how a producer or seller with market power can charge different buyers different prices for the same product. Perfect price discrimination involves setting individualised prices at every buyer's maximum willingness to pay. That means all the gains from trade go to the price-setter. Being able to price discriminate is a sure-fire route to huge profits.

In the OTC market, banks had been able to use information about their customers' characteristics, and specific customer orders, to set different prices and fees for different customers depending on their ability and willingness to pay. Seeing the techniques of pricing and control that banks had used in the OTC dark markets as well as in the mortgage securities market, quasi-banks like Enron transported those techniques to the energy market.

Enron's main electronic trading platform, EnronOnline, gave the company a big lead in technical capabilities and market share. It became a money tree, and there was a widespread belief that this was because EnronOnline's traders could offer different prices at the same time to different customers—an extreme, high-tech mode of price discrimination. (There was also a belief that energy traders, just like the banks in the OTC market, could front-run customer trades and otherwise use the information of customers and counterparties against them.)

Control was the key to profits on Wall Street and in energy

markets. The ability of traders and dealers to influence prices, and to use information to their own benefit, had allowed them to build money machines. Price discrimination in particular had been a marvellous prize in the OTC market and on platforms like EnronOnline. Was it possible, we asked, that the same prize was available in Australia's water market? And if so, what was the size of the prize?

We knew—just as the traders did—that the various types of farmers differed distinctly in their willingness to pay for water. In economic terms, the elasticity of their demand depended on what they grew and where they grew it. (Elasticity is a measure of the extent to which demand changes in response to a change in price. Demand for water is 'inelastic' because water is as an essential and non-substitutable farming input.)

The farmers we spoke to were aware of how the traders exploited this. 'Across the seasons and the drought cycle,' Peter Burke told us, 'different farmers have different willingness to pay. Some have signed three- to five-year contracts to grow crops, so they have to buy water no matter what. Some have permanent plantings such as almonds—there is a lot of money at stake in those trees, and the farmers can't just let them die when water is expensive. So they pay. And dairy cows, you can't just turn them off. So the dairy farmers pay, too. The brokers and traders know all this, and they use that knowledge in the market.'

'When the water's scarce,' Laurie Beer said, 'it isn't that hard for players to buy a lot of water at the price that's going at the time, so that you've got a lot of water up your sleeve and there's not much spare water in the market, and then you just jack the price up to whatever you feel you can get away with and put it on the market in dribs and drabs so it doesn't create any oversupply, and desperate people—like dairy farmers—they just buy it. They've really got no choice.'

'The investors wouldn't admit that they game the market,' Ben Dal Broi said, 'but how could they not? They work to the planting windows, like with cotton and rice for the end of October and November in my valley, if you want to give these crops a chance to achieve their yield potential. They know the water price is going to spike as farmers make the call on whether to plant or not. So if you are an investor, you'll be saying, "I want to hold it back as long as possible up to planting," to try and put as much pressure or desperation on farmers as possible.'

'The traders are capable guys,' Chris Brooks said. 'They analyse the market, and can see that water has inelastic demand. When a guy has got a crop and he needs some water, he will pay two or three hundred dollars per meg to keep it alive. And a guy who's got dairy cattle, he might pay four or five hundred to keep his cows alive. And a bloke who has invested in almond trees, that have taken him ten years to establish, he will pay $1,000 per megalitre because he can't have them die.

'So these trader blokes can come in, quite simply, quite legally, quite morally, and take very large positions. Now $40 million, let me tell you, that's chicken shit for the traders. They have that in the petty cash tin to jump on opportunities like this. And they would hold that water off the market and create a short and then sell it and get a 250 per cent return in six months.

'Don't blame the trader. He's doing his job. It's simply because the environment has been created for him to come into that market. And that could be changed at the stroke of a pen. So that's one very simple exercise. And then you get politicians saying, "Oh that doesn't pass the pub test, to take away a marketing opportunity for external investors."

'And they say they are concerned about compensation. Hang on, the big traders bought this water for nothing a few years ago. They've driven the price to record highs. What a good opportunity for them to get out now by changing the rules! Let them get out and leave the volume of water back in the hands of people who actually try to produce food for Australians.'

To pursue the golden opportunity of price discrimination, the most ambitious and cashed-up traders used their full set of tools, including their files on farmers, their arbitrage tactics and their water bots. They could buy and sell temporary water to bid up the price in the market. And they could buy long-term water rights then lease them to farmers at the level of their maximum willingness to pay. Or they could simply use their brute market force, and their brimming money tin, to buy up water and dribble it out at the monopoly price.

How to match price to 'willingness to pay': an insider's guide to monopoly pricing for dark water market cartels

STEP 1. Use bots to buy well-priced and undervalued water. The bots will ensure you get to it first.

STEP 2. Sit and wait.

STEP 3. Ramp up the price to a level at or above what you think is the maximum willingness to pay for farmers in the relevant region (or nearby connected ones) at this time of year and this point in the drought cycle. The detailed information you hold about the farmers in the region will assist you in this.

STEP 4. Hold a form of 'Dutch auction' (also known as a 'descending price auction') on the exchange or

> across a combination of on-exchange and off-exchange settings. The Dutch auction will allow you to pick off, from above, bids from purchasers who have the highest willingness to pay.
>
> Assisted by the artificial scarcity of the Holy Grail strategy, this 'auction' allows you to discover valuable information along the way about farmers' water needs and preferences. And it allows you to scoop up the available gains from trade.
>
> STEP 5. Repeat. Every time you do this, you will learn more and get better at it.

The traders used a wide variety of tactics: scraping, banking, leasing, hoarding, shuffling and bots. The water traders' smart bots could crawl across multiple trading platforms, which for farmers were not interoperable, but in traders' systems were seamlessly integrated. Thus equipped, the large traders could charge different prices at different times and to different farmers. They could even offer the same parcels of water on multiple exchanges at different prices simultaneously, thus achieving Enron-style price discrimination.

These monopolistic tactics in the water market were a natural consequence of two things coming together: traders' market power; and their detailed knowledge about farmers' water holdings and water needs. Price discrimination was the dividend from the traders' dual investments in information and power.

Implementation of the blended, fast–slow, megafarm strategy required sophisticated and coordinated trades across the whole Basin. Its centrepiece was a perfect, drought-driven arbitrage: buy water when it is cheap, and sell in times of greatest need and scarcity. The large traders could hoard or withhold water until there were peaks in demand, and then drip-feed it back into the market at monopoly-level prices.

'The crux of the issue,' farmer Rob McGavin told the ABC's *Signal* podcast, 'is the entitlements that are owned by non-irrigators. They should have an obligation to lease that allocation water or the temp water or the available water that comes out, in a timely manner to irrigators. And probably my biggest beef by miles is that non-irrigators can buy allocation or temp water and just hoard it and collect it and short the market of it. Every day that they're in the market bidding against irrigators means irrigators have got to pay more.'

'Investors will hold the water back as long as possible,' Ben Dal Broi said, 'in the expectation that farmers will over-plant and suffer a shortfall, or the summer is particularly hot and dry, and as such be willing to pay a premium to finish their crops.'

As Laurie Beer said, the traders were 'buying water rights from people desperate for the money, and selling it to people desperate for the water'.

The traders could do this within a year, across the seasons, or even between years, thanks to the market rules about carryover, which, perversely, the traders used to transform temporary, low-security water into a high-security water asset.

The profile of Basin water prices between 2010 and now shows the impact of the megafarms strategy. Prices of temporary water reached exceptional levels in 2007, but those spiky peaks were fleeting artefacts of extreme volatility in the market at that time. Since then, prices have steadily increased, and they are again approaching exceptionally high levels, but this time the peaks are much more consistent and more widely distributed.

As rural journalist Nikolai Beilharz reported on ABC *Country Hour* in November 2019, temporary water rights had settled at around $900 a megalitre, a level that was 'hugely higher than normal water prices that we've seen over recent years and there are questions about whether it could rise again as water demand for summer crops peaks'. Prices of permanent water were equally striking: '[O]ne of the things that seems to jump out,' Beilharz said, 'is just the seemingly inevitable climb of the value of permanent water. It's just going up and up and up.'

The potential to use market power in the water market is another reason the blended strategy was the Holy Grail. Tactics such as price discrimination unlocked enormous profits as traders would waste no upside: no surplus would be left 'on the table' in transactions with farmers.

CHAPTER 23
THE SIZE OF THE PRIZE

'For farmers,' Ben Dal Broi said, 'trying to plan how much crop they can plant in each season is a constant stress. They must weigh up the enormous complexity of how much water they have, how much they might be allocated, whether or not it might rain and how much, where the market might move before and after each planting window—to try to make up any shortfall. It has an impact on how much ground they prepare for a given crop, which is no small deal as the money outlaid on ground preparation, pre-plant fertilisers, seed, and the labour that goes with it can be eyewatering.

'In these situations, some farmers plant more crop than they have water budgeted for, in the gamble that the price of temporary water will come down when they need it to finish the crop. There have been occasions where some farmers have lost their complete profit buying expensive water to finish a crop—making them what I call "recreational farmers".

'The farmers who take the alternative approach and plant conservative areas lose out as they can limit their earning potential and vital cashflow should water prices be lower. That can be just as damaging as you need those good times to help support yourselves through the bad. The end product is that farming is less profitable, which makes it less resilient to absorb shocks such as droughts or crop failures, which makes life that much more uncertain for farms, and for the businesses and communities that depend on them.'

For the smart traders and water barons, the lucrative Holy Grail strategy delivered a degree of monopoly control over water prices: the traders and barons 'made' prices rather than 'taking' them. The water lords could use their market power to maximise profits, such as by selling water at different, profit-maximising prices to different customers and at different points in the drought cycle.

Before coming upon this strategy, the traders had the maths and the science to do this. After the Grail strategy, they had the market clout, too. The size of the prize was huge—much bigger than the water market itself. Market power offered the water barons the ability to set temporary and permanent water prices at the 'walk-away' level for farmers. At that level, farmers would earn no profits. They would be forced to their break-even point—even the large-scale cotton and rice growers—and all the available profit in the system would go to the water lords.

'Traders do have market power,' a Basin water trader told us. 'And what that means is that the water is priced at essentially the farmers' walk-away level.'

The ultimate prize, therefore, from $26 billion in tradable water rights was the total profit from all irrigated agriculture in the

Murray–Darling Basin. For the traders, the size of the prize over the next decade was the net present value of future profits from more than $240 billion in agricultural production.

'I'm not sure that many people realise,' Ben Dal Broi said, 'that if today's trends continue, our future irrigation farmers will be forced to rent the water they need as ownership concentrates to a smaller and smaller number of investors. Farmers will effectively be indentured tenants on their own land.'

Speaking in a 2019 documentary, dairy farmer Bart Doohan echoed this idea. 'With the big investors buying water,' he said, 'it's getting back to the old days of the landlords and the peasant farmers. And if we want to survive or stay into it, we are going to have to pay what the landlords deem the water is worth. Back to the Middle Ages.'

In quest of the Grail, the water Templars partnered up in various ways with agribusinesses, and they adopted a variety of different corporate and capital structures, but the underlying strategy was the same: use slow, cheap money to invest in visible, irrigated assets; and use fast, agile, hungry money to play the water market, largely in secret. In this way, the megafarms were the entree to the highest returns that the banks and traders could earn, reliably, year on year, with minimal or no risk. The results for the big traders were pure cream: Enron-style, hedge fund–level returns of 30 to 100 per cent per annum—or even more.

One group of traders earned an annual return from water trading of more than 150 per cent, and even higher returns have been cited. This meant the traders doubled or tripled their money each year by playing the Australian water market, without actually doing much with the water except cornering it, bidding it up and doling it out;

no genuine 'investing', and taking on little or no risk.

'In hindsight,' broker Desiree Button said, 'water trading has caused our region some problems. It's an open market, with supply and demand setting the trends. That has had some detrimental effects on certain parts of the Murray system, and probably the Murrumbidgee, too, to some degree. It has expanded irrigation downstream, and created wealth for those who accumulated rights. And it has obviously created significant wealth and opportunities for people who have absolutely no connection with agriculture, or Australia for that matter. It has made them vastly wealthy and, whether we like it or not, that is just the way of the world.'

The Templars could evade scrutiny by hiding their aggressive and ambitious trading within the needs and operations of their undemanding partners. And they could present their returns as the dividend from pro-social, pro-environmental investments in otherwise arid regions and country.

This strategy played out over seasons and even over years. According to Basin traders, it made the payoffs from intra-day arbitrage look like crumbs.

'With everyday arbitrage,' one trader said, 'we took the cream. But now, we take the milk, too.'

Public water accounting is unhelpfully vague, and so is private water accounting.

'The accounts of banks and traders are useless,' a broker said. 'No one can tell what is going on.'

From the companies' published accounts and overt corporate structures, it was almost impossible to see where one business ended and the other started. The traders adopted a variety of corporate forms and international structures that allowed them to pursue the blended strategy. The water barons favoured unlisted businesses and private corporate shells, which preserved their cloaks of secrecy and ambiguity. They regularly set up, closed down and renamed trading entities, which made tracing their activities and ownership extremely difficult.

(According to the fine print in the Victorian Water Register, publication of the names of entities that owned a high proportion of high-reliability water shares was limited to 'ownership in exactly the same name. Water shares owned by related parties have not been added up, as the Victorian Water Register does not collect information about company structures which would be needed to determine which companies are related to one another.')

Agribusiness players and water investors could hide profits or shift them offshore by using methods such as 'transfer pricing', which allowed them to make hefty payments between internal divisions supposedly as fees for the use of inputs, technology and brands. The inter-divisional structures were also helpful in blending and obscuring different types of water market activities.

The water barons' strategy was smart from a financial return point of view, and from the point of view of governance and scrutiny. And another reason the strategy was smart: it was future-proof. The strategy was robust, for example, to hypothetical policy changes, even ones that limited trading of entitlements to 'water in use' (i.e. only allowing water users to trade water).

The Holy Grail strategy combined multiple existing Wall Street tactics, but in its final form it was unique. It was only made possible by the scale of the traders' ambition, and the unique leverage that the market designers and regulators had inadvertently provided, allowing the traders not only to manipulate the market in a small way, but to game the market on a system-wide scale.

—

The water market was designed on the premise that no participant would have market power—that the water market as a whole would set prices, and every individual participant would be a 'price taker' rather than a 'price maker'. But in practice, cashed-up (and leveraged-up) traders were able to influence prices for their own gain. And they could do this without owning most or all of the Basin water rights.

This is an important point and a common area of misunderstanding. In the energy market, traders showed they did not need to own all of a type of energy or a class of derivatives to dominate the market for them. Enron controlled only about a quarter of the gas supply, for example, but it had effective control of the gas market.

In water, most of the available rights are seldom traded—this includes many rights held by megafarms and family farms. To substantially influence the market, all the traders had to do was substantially influence trading on any given day.

In their 2020 paper 'Treating water markets like stock markets', Constantin Seidl, Sarah Ann Wheeler and Alec Zuo from the University of Adelaide showed Basin water traders dominated the Australian market for forward water contracts, though their market share was 37 per cent. The researchers found that the traders primarily used the forward market for intertemporal arbitrage.

The Holy Grail strategy was aided by tactics that sent false price signals into the market, and tools that gave the traders a speed advantage and an information advantage. For players with deep pockets, executing this blended strategy was surprisingly easy. When the water price was low, buying cheap water was straightforward. And when the price was high, water was typically scarce, the market was as shallow as the Choke, and small interventions could have a big impact.

'They drip-feed small quantities of water to keep the price up,' farmer Peter Burke said. 'We see what they are doing—making big profits when water is expensive. Our only hope is that they get burned on the way down.'

But the traders didn't get burned. They had multiple layers of protection, should the market move against them. In particular, to guard against losses, the traders had the advantage of better information about market trends and weather; they were faster 'on the button', so they could be first in the queue to sell; and they could hedge their risks, such as by taking side bets that paid off if prices abruptly fell. The traders were utterly at home in the markets for derivatives and other hedging and risk-management products.

A small number of active players, therefore, could have an impact out of proportion to their total market share. And they could play the market with minimal cost and minimal risk. This was an alluring prospect, and it explained Wall Street's water enthusiasm. Through an evolutionary process, water traders had arrived at a dominant strategy: a lazy portfolio of industrial, pension-funded, mass agriculture; plus fast-paced, high-octane, high-return water trading. The equation was compelling. Slow money plus fast money created long-term, stable returns, plus supernormal, low-risk payoffs. Definitely worth getting out of bed for.

CHAPTER 24
THE SUCKING SOUND

Large, industrial farms below the Barmah Choke—vast fields of olives and almonds and other novel crops—are a highly noticeable feature of Australia's new water world. These megafarms are largely on previously unirrigated land. A Basin farmer, interviewed in 2007 as part of the AgriFutures study, explained some of the advantages for megafarms in planting on previously unirrigated land downriver.

'You can't put under-vine sprinklers on old trellises that are too wide,' the farmer said, 'the plantings were all wrong, and there is no point putting a Rolls Royce irrigation system on clapped-out old plantings. Also, they manage their own water infrastructure rather than draw water from the water authority's system. The difference is in the pump district; the water costs are higher because the growers have to contribute to the infrastructure and the pumping charges. But the cost for private diverters out of the river system is a heck of a lot less.'

Another Basin farmer noted how the megafarms consumed a lot of water and put pressure on the river and associated delivery systems, especially during summer and in times of drought.

'Water is brought in from other districts or from sleeper licences, which places stress on the river systems. With the drought as it is at the moment, it's illogical the amount of water that is being used and is still brought into the area to be used on totally unirrigated properties. In the heat of summer it is going to be a problem for everyone.'

People in Basin communities shared a variety of concerns about the developments, including that they imposed whole-of-system costs and they reflected unfair advantages over family farmers. There was widespread scepticism about the efficiency and profitability of the downriver ventures.

'One of the things that I really have a beef about,' an agribusiness consultant said, 'is this talk about sending water to a more efficient use. Who's buying the water? Almonds, olives and grapes? What are we going to do next?...All we've done is shifted water from basically productive family farming to tax-driven enterprises creating gluts for worthless products. I don't see any sensible planning in the system that allows that to happen. Is that good management? It certainly goes under the guise of good management, but it's ridiculous.'

Other farmers expressed the same concerns about the wisdom and economics of the megafarms below the Choke.

'This huge place going up at Boundary Bend, on the sand, if it was in the US, no way they would let them do it. Because they are putting on 14 to 15 megs a hectare to grow almonds, and that is way above any other use. And everybody is saying exactly the same thing. In the US it would not happen because it is not efficient: all they are doing is watering sand. So don't give me this argument that it is best practice. If you want to create a system that is sustainable long-term and ride the peaks and troughs, that system has to be designed around your generic prime lambs, dairy, cropping scenarios. I don't think you can build a system around fads and fashions.'

'I can't understand why water is moving to investment on the Lower Murray,' a Basin businessperson said. 'Why isn't the investment coming to the water? They create a bit of a false economy: they can pay whatever they like for water. Temporary water is currently $650 a megalitre. It would be interesting to see what it would be if the [managed investment schemes] didn't exist.'

A Basin farmer saw perverse market forces behind the movement of water rights downriver. 'There is a lot of water being bought to grow grapes and olives, but I'd love to see where their markets are. I wonder how much is being bought by the superannuation funds and the like...Show me where the wine market is. I can go and buy any number of $2 bottles of cleanskin wines. You can't tell me they are making any money on that...We are not going to wash in it. There must be a ceiling in wine production, and I think we are a million miles above that ceiling. And the same with olive oil: I don't have a bottle of that in the fridge and drink that every night.'

A view, widely expressed, was that tax minimisation schemes and other rorts and biases were accelerating the movement of water to the new developments.

'My particular beef,' a Tatura agribusiness owner said, 'is the seemingly clear government policy of promoting irrigation development (often through managed investment schemes) downstream of established irrigation areas. These developments take agriculture to areas where there is minimal infrastructure, processing or community development. The tax breaks offered make them attractive against mainstream agriculture. In other words, we are exporting our water to areas which provide tax benefits—which doesn't necessarily equate to efficient agriculture—at the cost of previous government expenditure at the regional level.'

A local water trader was concerned about the long-term effects of the flow of water downriver. 'I'm really cheesed off about the way it's going. The fact is that it is getting wasted downstream and causing us huge amounts of cost and angst. We've got to put in infrastructure, build a new Murray to get more water down there. For what? They'll all be bankrupt in three years' time. What are we going to do then?'

With the passing of time, the concerns of Basin communities can now be seen with more clarity. The concerns have in large part been vindicated. Despite what a generation of water regulators and policy people argued, we now know the megafarm investments were not driven by the irrigation value of water and the market for agricultural products. Instead, they were driven by capital markets and water trading.

The megafarms were presented as highly water efficient, but in fact they were not. The industrial-scale farms had a lot in common with other failed schemes in the Basin—such as timber plantations and emu farms—which were motivated by taxation and property speculation rather than agriculture, and which came and went during the 1980s, 1990s and 2000s. The megafarms were the latest bubble in the Basin, as broker Desiree Button explained.

'I'm a big believer, without being a socialist, that Australia needs and deserves to have the very best staple food produced—and that is a combination of agriculture. The bulk of the staples are still grown above the Choke. The big traders and big corporate investment—their bank balances would wipe us out above the Choke if they were allowed to continue to grow at the rate they have.

'A lot of these nuts, vineyards, orchards of all descriptions, whether it's olives or almonds or walnuts, macadamias—once they get to around six to ten years and beyond, they use as much water as a rice crop or anything else. They use 10 to 14 megs to the hectare easily. They have marketed themselves very well as being highly efficient producers of food. But at the end of the day, most plants, especially if you are watering all year round, like trees, and as they grow, their water demand substantially increases. There is going to be a day of reckoning.'

'Hear that giant sucking sound?' another water broker said. 'It's the sound of the market sending water below the Choke.'

The sucking sound of water rushing from the top of the Basin catchments to properties far downriver was a broken promise for dairy farmers and long-term foodbowl regions such as Kyabram and Katandra West in Victoria.

'The concern I have,' a financial adviser said in the 2007 study, 'is the dramatic shifts in water moving into different regions. Areas that used to have significant irrigation infrastructure are almost being disbanded. Katandra West is a good example: there's been a lot of trade out of that area and, historically, this was an excellent area for dairy farmers.'

The novel crops downstream are often grouped in a common category with traditional upstream products such as milk and fruit. But there are important differences between the food produced upstream on predominantly family-owned farms, and that produced downstream in industrial megafarms.

One key difference is in how the family farms and the megafarms are financed. Most family farms are funded from retained profits (from prior good years) and current revenue. But the megafarms are funded by external capital—and on a doubtful promise. The banks and traders helped to solicit and direct the lazy capital that supported the megafarms, just as they solicited borrowers for the 'originate to distribute' mortgages model.

The vast orchards funded by outside investors shared a problem with securitised mortgages. Once the money was handed over, the recipients had little incentive to make sure the underlying asset was a good investment. Compared to multi-generation family farms, the managers of investment scheme–funded farms had cheaper capital and weaker incentives to ensure the land was used well and the returns were delivered as promised. The experience with securitised mortgages and the global financial crisis showed we can't rely on capital market disciplines to drive good behaviour; and that fact is again being proven, this time in Australia's water market.

This boils down to an issue of 'risk allocation', and specifically a question of who might win and who might lose as a result of the farming investment decisions. With a family farm, the people making the decisions—such as whether and at what scale to plant a particular crop—bore the upside and downside risks associated with those decisions. The economic costs and benefits of good and bad decisions mainly rested with the family.

With pension-funded industrial farms, however, the downside risks were mainly borne by the superannuation account holders and beneficiaries, but those people were far removed from decisions about where and how to invest. And a large part of the benefits were earned not by the account holders and beneficiaries, but by the large traders in the water market.

There were other forces, too, that helped send capital towards the megafarms and away from family farms. Apart from the megafarms' tax advantages and lower capital costs, they had much better access to capital markets. The megafarms could raise money from the share market, superannuation funds, pension funds, hedge funds and other sources, but family farms could not borrow against their water assets, because they were not legally 'property'.

Flaws in the water market design made the megafarms strategy more viable, and helped accelerate the unexpected westward movement of water to arid country below the Choke. Those flaws included unaccounted-for water losses; misallocated environmental costs; anomalies in irrigation scheme costs; and the soft regulatory framework that allowed predatory trading, including in its most extreme, highly developed, monopolistic form.

Irrigation communities upriver were lumbered with the costs of legacy infrastructure and stranded assets.

'I grew up here,' farmer Chris Brooks said, 'and went to school at Assumption College. I worked in our family business, which was buying grain from locals. We built storage along the Murray River, all the way from Albury to Mildura. We exported a lot of grain and we sold our business to Glencore, which is a big international company, and I was the Australian CEO of Glencore for ten years. We were probably one of the biggest exporters of grain from Australia for that decade, into all the Asian and Middle Eastern markets. I retired at age 50, about ten years ago.

'In 2012, I came home to buy the family farm and enjoy life. It was a well-structured farm of about 6,000 acres with 23 centre pivots. And we owned a lot of water, 5,000 megs. Unfortunately, that was also when the Murray–Darling Basin Authority was formed. It took me a couple of years to work out that this massive water licence I'd paid for wasn't actually delivering, and it was affecting my farming model pretty substantially. So I got looking for the water and asking a lot of questions, and it's just been a never-ending search—a lot of hollow logs—as to where it went and why.

'I was a commodity trader. I don't believe in protectionism or socialism. I'm a very commercial operator. When the government allocated water to farms, that was a great thing. When they disconnected it from the land, that was a stupid thing. But they never deliberately disconnected it from the region. When you've got infrastructure here, like we have with Murray Irrigation, the valleys are set up in such a way that there's a storage, there's the resource, there's a network of channels and there's a fixed cost to run it.

'If there is a million megalitres of water and it costs $10 million, they levy the growers $10 each megalitre and that works perfectly. Suddenly you take half the water away—but those costs don't run away. It's still $10 million to maintain it and the cost to the grower goes to $20 a megalitre. So therein lies an obvious problem when you have white-anting of water from regions.'

Some commentators have argued that there should be more extensive use of intra-Basin 'exchange rates' to reflect the different costs and values associated with water in different valleys and regions. Due to transfer losses and environmental impacts, for example, one gigalitre of water delivered below the Choke was not the same as one gigalitre above. Without some form of appropriate price adjustment, people buying water downriver were being subsidised because they didn't pay the full cost of water losses and impacts along the way. The farmers we spoke to confirmed this.

'By our calculations,' Chris Brooks said, 'the losses through the Choke are 37 per cent. And we're the ones who pay for it.'

'All the conveyance losses have to be accounted for,' Shelley Scoullar said, 'before Murray General Security get an allocation. So basically, we're underwriting the transfer of water to the South Australian border, and when the Darling's offline, we've got to deliver their share and the conveyance losses as well.'

These forces combined to make the blended Holy Grail strategy more viable and easier to implement. The 'competition' that was sucking water and capital from dairy farms and foodbowl communities was not farming competition, or an instance of 'allocative efficiency'. The large, industrial monoculture farms were generating low returns, but the people orchestrating those crop investments were making supernormal profits, many of them from an enormous side hustle through the water market itself.

The most compelling images of the market failures that led to the global financial crisis were pictures of overbuilding: vast fields of newly constructed houses—in California, Texas and Florida—that were unoccupied and unsaleable. The causes of the overbuilding are well established: poorly designed financial markets, plus a mistaken idea that the risks associated with securitised mortgages (and bets on those mortgages) were well understood and well managed. At bottom, there was an idea, based on ideology rather than evidence, that abstract market forces would protect consumers—and the system as a whole—from the dangers of unbridled greed.

In Australia, the most compelling images of our market failure will probably feature a different kind of misdirected investment. Not overbuilding but overplanting. The images will likely be pictures of abandoned fields of almonds and olive groves—as they prove to be thirsty and uneconomic, just a by-product of dark-market gouging. The counterpoints to those hypothetical future images are real current ones—of abandoned dairy farms, fire-sale equipment auctions, derelict sporting fields and other signs of economic retreat in formerly thriving foodbowl towns.

—

Farmers could see evidence of the Holy Grail strategy being put into practice. They could see, for example, that the market was sending water in strange directions, and they could feel the impact on prices.

'Water has gone from hundreds of dollars per megalitre to thousands of dollars,' Lachlan Marshall said. 'The amount of irrigation that happens [in his up-river district] is probably less than half of what it was. Someone is making a killing, but not the farmers. Water trading has destroyed our community.'

'For corporate farming to grow,' an irrigator said in 2007, 'it has to get water from traditional irrigators. This benefits only the corporate investors, and destroys family farms.'

'As you know,' Chris Brooks told us, 'this downstream demand is driven by the transfer of water and the trading that is allowed to continue down there. If you go back through the processes of the formation of the Basin Plan and the full recommendations of National Water Initiative—they stipulate that there was restricted volumes that were to be traded inter-valley, and in any case any losses incurred in trading water downstream were always for the account of the downstream user.

'Unfortunately, they take all of the conveyance and all of the losses only from the New South Wales Murray General, and that's got to stop. We are not responsible for the losses incurred by South Australians or almond growers or the environment or any other bastard downstream who wants to take the water. First of all, it should be complying with the National Water Initiative and, secondly, it should be their responsibility to pay their share of the losses.

'We would actually argue that we don't incur any losses. We are at the foot of the mountains. I can see them out my window and that's where the dam was built and that's why all this infrastructure is here. And that's why all the rice mills and the dairy plants and intensive livestock breeding is here. Because it was to drought-proof the region. To relocate that water to some corporate almond farmer owned by a foreign national is just insane. It's just absolutely insane.'

CHAPTER 25

A CATASTROPHIC ERROR

In the water market underground, the idea that traders could exert market power had long been the subject of conspiracy theories and bogeyman stories. But regulators swore blind that no water market participants could exert market power to 'short' the market and price discriminate. On that point they were very consistent, and it underpinned their interventions in policy development, and their assurances to Basin farmers.

But we are more than confident that, if and when a future audit or inquiry looks at this aspect of Australia's water market experiment, those regulators will have a lot of explaining to do. We are confident, too, that they will not be alone in the hot seat. If you listen very carefully right now, you can hear the haptic hum and crackle of traders and agribusinesses shredding documents, deleting files and destroying USB sticks that contain evidence of their coordination and partnering.

—

Right back at the beginning of the water market, in early milestones such as the National Water Initiative and the creation of tradable rights, the intention (and the rhetoric) of politicians, policymakers and regulators was that the unbundled water rights would flow to their highest and best uses. But through the market design, what actually happened was that the rights flowed to where there was the greatest prospect of a financial return.

This is our central finding. Policymakers had made a critical assumption that both of those things would be the same—that the highest and best economic use would also provide the highest financial return—but they are emphatically not. This is the basic, catastrophic mistake at the heart of our water market.

The water barons have played the market in ways that are permitted by the lax rules and protected by the ideological belief that no water market participant would be able to wield market power. There is a grave disconnect between the water market design and the reality. The market designers' mistake—confusing economic returns for financial ones—has driven the market, the mix of agricultural output, and the drain of capital from foodbowl communities.

There is another important disconnect here, too. The water market was meant to be about place and country. The market, so the policy alchemists said, would move water from one farm or region to another, where the water would be best used. But the reality of today's water market is very different. The people who drive the market do not have a state or region or even a country. Their interest in water is estranged from place.

The big traders operate in an unreal, footloose, almost virtual world. The value driving the market is the value of water in their hands. That value pivots on how the barons can use water in the market itself to earn supernormal returns. It does not depend on a particular, physical place. From their point of view, they might just as well be dealing in Brazilian soybeans or bets on the World Series.

The water reforms unbundled water from land, but they were not supposed to separate water from geography. The disconnected, ethereal character of the modern water market is another sense in which the original goals and concepts of the market have been perverted.

'I'm concerned,' farmer Ben Dal Broi said, 'that an unintended consequence of deregulation of water ownership is that producers relying upon water will eventually lose control of it. Since deregulation there are others that farmers now have to compete with to own the water that they use. These include retired farmers that can sell the land and keep their entitlements to boost their retirement incomes.

'Then you've got your corporates and bankers and big investors. I remember at an auction that they bought a stack of water and everyone's thinking, "Why are they paying through the nose for this water," around $5,000 a meg for high security and they were just absorbing it all. They had *deep pockets*. And they had a grander plan as well: get up a big enough parcel and then you've got a saleable item for a bigger corporate, probably one from overseas.

'I do worry about the impact on the water market from big corporates that are putting in mega-hectares of permanent planting such as almonds, particularly down lower in the Basin. These include pension funds from the US and Canada. Somebody asked me, why aren't the Australian super funds doing it too? I said, well, Australian super funds need to keep money ready in case their members ask to be paid their super as a lump sum during retirement. So they've got to have relatively liquid assets. Whereas foreign, state-owned pension funds paying only pensions can invest patient long-term capital. No doubt they're looking at our market and think that water is good value. Water is a sexy thing globally to be investing in at the moment.

'But this probably introduces you to another problem—the inter-generational ticking time bomb, as I call it. The water market is going to concentrate ownership with fewer and fewer holders, giving them enormous market power over producers. I'll explain this ticking time bomb by outlining my situation.

'My parents and I farm together near Griffith in New South Wales and I have two siblings. We have a succession plan in place that is designed to distribute my parents' estate after they are gone, and for my part to help me continue farming. The plan in a nutshell is that my parents distribute profits to each of us when they are able. My share is to build equity in the farms.

'So when that sad time comes, I should in theory have a larger percentage of their estate than one-third and be in a better position should I choose to buy my siblings out. However, remember we're not talking of only the land but also their share of the water, both of which I would need if I were to realistically continue farming. My siblings would be entitled to market rates for their share of water because that's what's fair.

'I need the water to run my business and I could come to an arrangement to lease it from them, but they might exercise their right to sell and put their money to use elsewhere. If that happens and I want to buy the entitlements, I'll have to find a lot of money, so it's going to be a big hit to my business equity. In all likelihood, it just won't be possible as I'll be competing on the market against people with deeper pockets. These will likely be the corporates, investors, super and pension funds.

'So imagine there are lots of different family farmers across all irrigation industries in the same position. Over time, the son or daughter who chooses to stay will need to buy out the other siblings. If there are lots of siblings then the problem will only be harder. Each time an estate is divided, water will likely end up on the market to be bought by investors. Follow the bouncing ball and in time very few farmers will own the water they use, and they will have to pay dearly for the use of it.'

We thought now was a good time to take stock. We went back over what we'd learnt from listening and asking questions and turning over rocks.

Governments and regulators had encouraged farmers to enter the water market. They'd also invited banks and other non-farming traders to enter. That decision, however, was based on stale information. People had formed their everyday understandings of the words 'bank' and 'finance' in the old era of balance-sheet banking, when banks invested in home loans just like farmers invested in crops. But by the time of the global financial crisis, those days were long gone. Once inside the water market, the banks and traders used Wall Street techniques (and Wall Street ethics) to play the market.

They introduced digital trading techniques, aggressive arbitrage tactics and market manipulation. There were few controls to prevent conflicts of interest, such as those faced by brokers. Before entering the Australian water market, Wall Street traders and hedge funds had worked out how to corner markets, especially the 'dark markets' for over-the-counter swaps and derivatives, where they could trade on their own account, and use the information of customers and counterparties against them. The payoffs were huge, including secret margins and commissions, but the risks were low. This was a dark-market magic pudding.

In the late 1990s and early 2000s, American traders had translated those dark-market techniques into the market for energy. And in the 2000s and 2010s, the techniques reached the market for water. Through a deliberate decision of our water market designers ('the alchemists'), the banks and hedge funds and other outside investors were given a green light to play on a large scale. The alchemists minimised restrictions on the market, and this included stripping away protections that were otherwise available to investors and customers in Australian financial markets.

The Wall Street traders and hedge funds invested in information about farmers and irrigators, and in computerised trading and bots, to get an advantage over other market participants. The water market traders earned high returns, out of proportion to the risks they took—just like they had in the over-the-counter derivatives market. Brokers, too, earned high returns, including from conduct that would be banned in most Australian financial markets.

Well-resourced traders thrived in the lax regulation and high complexity of the water market. The complexity was a picnic and a cloak. Some traders claimed the disjointed and dysfunctional nature of the water market was a problem, but this was disingenuous: they thrived on the dysfunction. If the markets were smooth and transparent, the traders would not be able to make such easy profits. And if the market wasn't so foggy, more people outside the market would see what happened in the dark.

The most lucrative strategy of all involved shorting the market of water, creating artificial scarcity, and trading across seasons and years to maximise profits from arbitrage. The ability to do this—which reflects a culpable error at the heart of the water market design—drove the movement of water rights and the profile of Basin agriculture. Water moved in the opposite direction to where the alchemists first expected—and it moved in that perverse direction because of their mistake.

Due to weak oversight, the traders were able to align their efforts and coordinate their price-making trades. Coordination magnified their potential market power. The returns available for the water Templars were larger than the water market itself. In the limit, the size of the prize was the gross return from agricultural production in the Basin. (And this was not constrained by the current total return. The Templars could dream of a larger return if the water market became a catalyst for wholesale industrialisation of Basin agriculture.)

Price discrimination is a classic strategy of cartels. The viability of price discrimination in the water market is further confirmation that water trading is a dark market. Price discrimination in the water market was the natural consequence of two things coming together: the ability of big traders to exert market power, and the ability of those traders to know what farmers intended to do in the market. All those early investments in gathering information paid off in spades.

The Holy Grail strategy transformed the geography of Basin agriculture. For banks and hedge funds and other active traders, the lazy-money farms were a stable foothold and a long-term licence to operate on a large scale in the much more lucrative casino part of the water market. The industrial farms gave the traders extra heft, extra influence, and a compelling story to wrap around their dark market activities.

The traders didn't talk publicly about what they were doing, and would publicly deny they could manipulate the market or otherwise play it for profit. The market regulators refused to acknowledge that big traders could price discriminate and otherwise use market power. But insiders knew the Holy Grail strategy was being implemented, and was extremely lucrative. The Wall Street techniques and strategies were an open secret among the financially savvy brokers and market operators, and they affected the daily reality for farmers who sought to use the water market for farming.

If governments don't act quickly, Basin agriculture will become a kind of irrigated Ponzi scheme, with over-investment in unsustainable crops and megafarms. The rural impacts of the failed market design are not easily reversed.

The loss of family farms is asymmetric: it suffers from the problem of 'hysteresis', which is a type of 'path dependence'. The farms that are closing have been built over generations, and the people who are leaving are people with generations of knowledge. The loss of these farms cannot be quickly restored or reversed. They cannot be rebuilt overnight. If the current Wild West water market is just a temporary, experimental phase of development towards a more balanced and effective market, the time period will still be too long to save these farms. They won't survive the experiment.

The water market as a whole is obscured by ambiguity and dissimulation. Regulators speak without precision about 'efficient markets' and market failure. Academics and consultants, too, get lost in a linguistic bog. To the extent that bankers and traders report on their Basin activities, the reports are full of doublespeak. In their investor collateral, for example, and in their submissions to governments and panels of inquiry, they claim to be 'helping' farmers and irrigators by bringing liquidity and otherwise 'supporting' their irrigation 'partners'. But the reality is quite different. Here is a water market decoder:

- **Market designers:** 'The market was designed to harness competition and to elicit true information about farmers' needs and valuations.' Translated: 'With no central market and no single "price", the market was almost designed to be played—to the extent it was designed at all. It is a paradise for arbitrageurs and volatility traders.'
- **Wall Street bankers and hedge funds:** 'We are "making the market" in order to help irrigators implement their plans.' Translated: 'We are playing the market, hard, and we are eager to continue. We are willing to lie to cover up what we are doing and how much money we are making.'
- **Officials:** 'The market is functioning well. Water is flowing to the highest and best uses.' Translated: 'Water is flowing to where the banks and megafarms can make the biggest returns—including huge trading profits from the blended Holy Grail strategy. The mega-returns are disconnected from direct investment in agriculture.'
- **Regulators:** 'There is no scope to exert monopoly power in the water market.' Translated: 'Large participants probably do manipulate the market and exert monopoly power, but we don't have the tools or the appetite to test our *a priori* presumption that no private players have cornered the market for water.'
- **Politicians:** 'We are assured by the MDBA, the ACCC and other advisers and agencies that the

market is functioning as intended.' Translated: 'The whole thing is extremely complex and probably a mess. We know something has gone badly wrong, and we don't want to be left holding the bag.'

PART V: FIXING THE MARKET

CHAPTER 26

WHAT THE FUCK JUST HAPPENED?

In March 2020 we interviewed former MP Tony Windsor about political decision-making and bipartisanship. We asked him about his decision to shift from the New South Wales seat of Tamworth to the federal seat of New England.

'I had the safest seat in New South Wales at the time,' he said. 'I had 70 or 80 per cent of the two-party preferred vote. I used to jokingly say the move was about the money, because I think federal parliamentarians got $10 a week more! But the main reason I had a go at the federal seat was about competition policy. I kept running into a whole range of policy issues where the state or local government was affected by the Commonwealth government.

'In a non-democratic way, the Commonwealth through the Council of Australian Governments process was able to circumvent their own electorates. The state governments coming to an agreement with the Commonwealth virtually bypassed the state parliaments in terms of whatever the issue was, on the basis that it was in the national interest or it was this or that. Particularly for country people, they were getting fenced out of the political equation, because of the competition policy rules that were put in place.

'It wasn't a very glamorous reason for going federal. Some would ask, did I have great success in making a difference? Well, probably not, in terms of changing competition policy, but I think what we've seen just in recent years is that a lot of things that were put in place because of the competition policy rules have failed. Not just for regional people, but more broadly as well.'

In major public-policy interventions, it often takes years of sitting and germinating before the benefits and costs become clear—especially the perverse and unintended ones. We have seen this in aged care, for example, and in the privatisation and deregulation of electricity assets. We are also seeing it in the design of our water markets.

The error at the heart of the water market has been devastating for rural communities. Traders' supernormal returns from hyperactive arbitrage and their Holy Grail strategy accelerated the drain of capital and wealth from foodbowl towns. Noticeably less money was spent in the main streets of Moulamein, Echuca, Numurkah, Nathalia, Narrandera, Shepparton, Deniliquin, Griffith and Wagga Wagga. There were fewer school enrolments, fewer children's parties, fewer people in the footy club or the netball club.

'I was a farmer in Deniliquin,' Shelley Scoullar said. 'I grew up on a mixed farm where rice was the main enterprise. I went off and did a biotechnology degree, but decided I didn't want to be stuck in a lab, so I became a science teacher in Adelaide, then moved over to special education.

'Fate sort of happened and we ended up buying the farm across the road from Mum and Dad, in 2006. My husband worked off-farm at SunRice, doing the milling and the manufacturing side. I was running the farm with Dad, which was just awesome, the best thing I have ever done in my life. And then things went quite "not so good".

'When I was growing up, Deni was a thriving town. There were ten netball courts playing all day on a Saturday, the soccer fields going flat out. Now, there's not even a netball competition on a Saturday. So that's what's happened to our particular town.'

'I've moved a lot of permanent water out of these [irrigation] districts,' a water trader told the 2007 AgriFutures study. 'I prefer to see water stay in a district because of the social impact that it will have, and is having, on communities.'

'We need to pull it back somewhere in between,' farmer Laurie Beer said, 'so people can still run irrigated farms. The farms in this area were designed around irrigation. That's what people probably don't realise. Irrigation isn't just a supplement to us. It is the way our farms were designed. We're only relatively small farms and we're designed to get an irrigation allocation every year. We've got fairly unreliable rainfall, but by being able to do two or three waterings or whatever a year, we can turn a marginal crop into a good crop. But if you haven't got the water, you can't do that.

'So the only way is to turn them back into big land areas, back to landholdings that might be five, ten thousand acres, which is what they were before the irrigation system came through. If that happens, the towns and everything else will suffer. There won't be anywhere near the population living on the same area of land.

'People think irrigation water is just something like the cream on the cake, but it's not. It is absolutely what our farms were designed to be, irrigation farms. We need that irrigation water just to remain viable. As far as I'm concerned I could sell my water and live off the proceeds, live out my days, but it won't help the town or it won't help my children to make a living. And I won't be producing clean food for Australians and the rest of the world to eat.'

The impacts of the water market failure played out in commercial ways, such as in loan defaults, foreclosures and fire-sales of farming equipment; in family ways, such as inter-generational conflict, and anxiety about leaving a legacy; and in personal ways, such as depression, addiction and suicide.

Community leaders expressed concern about mental health in the Basin. An agribusiness representative told the AgriFutures study: 'Because of the drought, our workers become like counsellors. I actually had one of my men come back in tears and said, "You need to go out and see this bloke. He's talking about taking his own life." We went out a couple of times and just listened, and we got him through it. I think it will happen more. There are a lot of people who have done up dairies and have to buy more water and they can't and this is going to ruin their business. And it's just the price of water. I'd just like to see the water stay in the area and be capped.'

To fund their own investment, farmers relied primarily on retained earnings, augmented by borrowing from banks. But with no profits to retain, some farmers found themselves in a downward spiral.

A Basin agronomist said of Boort in Victoria: 'Permanent trade within the district is probably not an issue. But as soon as it leaves the district, every other irrigator has to pick up the bill for the infrastructure and the service. So you have the same cost spread across fewer people. It's probably going to get worse...In an area like this, if farmers aren't making money from their water and they have debt, the obvious thing is to sell the permanent water to get rid of the debt and trade on temporary water. But it's a continual downward cycle. The whole district could be based on temporary...The security of the area is at severe risk.'

Time after time, in the development and implementation of Basin water policy, government agencies and regulators painted foodbowl farmers as whingers, sore losers and troublemakers. Some adjustment costs were inevitable, the regulators and technocrats argued, as water moved to its most efficient use, but the movement was always going to be for the greater good. Among farmers, they promised, there would be as many winners as losers.

That familiar line, sadly, was a false promise. Dairy farmers in north-central and north-eastern Victoria have been hit especially hard. 'General security' farmers in southern New South Wales were also hammered. General security entitlements have seen a big drop in yields. In the hierarchical allocation system, these farmers bear much of the climate risk in the system—and they have borne much of the impact of the false scarcity caused by the water lords.

Here are some of the words we heard people use to describe Australia's irrigation water policy: naive, insane, ridiculous, broken, rooted. It was an epic fail. The policy weakened Australia's food supply chain at a time when the international opportunities for Australian food were greatest. The economic impacts have been felt in lost production and lost economic value.

'Basically,' farmer Chris Brooks said, 'the rule of thumb is that there is about $1.6 billion or $1.7 billion in direct agricultural production in this region, and there is a multiple of four in all the value-adding and associated activities. This region is all about dairy and rice and fruit, and everything is value-added and packed. It leaves the area in a bag, a carton or a box.

'We have 850,000 megs. In the case of the Murrumbidgee, they have double that, they have about 1.6 million megs—but they grow higher value produce, where we grow rice, the cotton, the almonds and walnuts, and they have more fruit and more chickens, more of the higher value stuff. Just doubling it, it's in the $12 billion range.

'And if you come back to the Victorian side, taking the entire Goulburn–Murray valley of Victoria. It's not only a bigger volume of water, but it's higher value because of the very intensive food processing. But using just the same analogy of, you know, dollars per megalitre—it's over $7 billion. So the three valleys, I'm saying, are $25 billion worth of economic value.'

Some of the biggest impacts of the water market failure are being felt in overt changes to rural culture. Apart from the loss of income and capital, there has been a loss of trust. The market is driving arbitrary wins and losses in rural communities. This, combined with the secrecy and falsehoods that surround the water market, is corrosive. (The market has had a toxic effect, too, on Basin politics.)

The old ways and the old ideas of how to make money, how to select crops, how to plan for the future: all these have been transformed. Farmers who cling to old patterns and ideas are floundering. Far removed from the early modes of water trading, the water market has been financialised and corporatised. Most jarring to farming communities: the new market and the new culture reward sedentary greed rather than agricultural ingenuity and effort.

In the old days, before water trading, the most hated rural bullies came from the usual places. Now, they are the water traders. Farmers have become much cagier about whom they talk to and what they can safely reveal about their needs and actions. They are permanently wary of sharing information that could be used against them in the market.

Many farmers—and even some brokers and traders—don't like what has happened in water. They think the market has gone in the wrong direction, with big costs for people who can't keep up. Many farmers, including Peter Burke and Laurie Beer, say they are doing their best to live with it, while hoping for beneficial reform.

The way the water market has evolved—into a kind of hydrological casino—is extremely awkward for the water brokers who live in rural communities and who work day-to-day with farmers and irrigators. The brokers see the worst of the market behaviours and the arbitrary transfers of wealth. In the market design, brokers were meant to protect farmers from the worst tendencies of a financialised water market. But in practice, they provide scant protection, and in some ways they actually make things worse.

Traders, too, have a lot to answer for. But a major trader sought to justify how he had harvested profit from the Basin by playing the market.

'If we don't do it,' he said, 'the larger and more switched-on farmers will, such as the big cotton growers in the north.'

We didn't accept that argument. There is a big difference between farmers bidding up water, and traders bidding it up. When farmers compete with farmers in the market, the profit stays in the Basin and it stays in agriculture. Also, the farmers only bid water up to a level where they can still earn normal profits. So the farmer-on-farmer water price level still includes a margin for farming profit. But with external traders driving the market, the water prices are set at the farmers' 'walk-away' level. There is no profit in the whole system, except what the traders receive.

Foreign involvement added an important dimension, as farmer and barrister Robyn Wheeler explained.

'International transactions are very, very hard to trace. That's why multimillionaires are involved. I just find all that quite sickening when you've got farmers around here, who are putting bread and beef, and lamb and rice—and, in Griffith, lettuces and wine and oranges—on the table for Australians to eat, whether they're rich, poor, ugly or indifferent—and going broke and committing suicide because they have got zero water allocation. Where they were told—when that legislation came in—it wouldn't affect the water availability. That was what they were told. That is what was promised. And that is just a lie.'

CHAPTER 27
PLANS ON FIRE

On their height-adjustable desks, alongside their Newton's cradles and Foreverspin tops, water traders displayed trophies: tombstone mementoes of especially profitable deals, plus framed clippings from the *Weekly Times* and *Deniliquin Pastoral Times*—souvenirs from when journalists got close to understanding their activities. Other artefacts, too, were prized, including charred copies of MDBA reports.

In 2010, there was a strong sense in Basin communities that regulators, officials and politicians did not understand the real impacts of water policy and water trading. Public consultation on the Basin Plan became a lightning rod for their frustration with the metropolitan policymakers, who were seen as meddlers and wreckers—a distant cadre of middle-class anarchists.

Farmer Laurie Beer expressed what had become a common feeling.

'As far as I'm concerned,' he said, 'the Murray–Darling Basin Plan is nothing but a scam. It's not a plan at all. It's just a scam. A total disaster. That's my view. I believe it's been written to a preconceived agenda.'

Farmer and barrister Robyn Wheeler understood what rural communities felt about the Plan, and in 2010 she played a part in helping farmers express their views.

'The reason the farmers were so irate,' she said, 'was that the Plan is something like seven volumes, and the first volume was released, really and truly, 24 hours before the Authority started its, in inverted commas, "consultation" meetings with farmers. And the first volume was I don't know how many pages thick. I downloaded it and printed it. You know, my clerk was printing it for half a day, for this first volume, and I started to read it.

'Now I'm a barrister and I read complex documents, and I went, "Oh my God." I looked at it. It had all these findings, scientific findings upon which it was based. None of which were in that volume. So they were going to be produced at a later time. But therefore within 24 hours you were supposed to digest this document, have a reasonable point of view and have a rational discussion about it with the Authority.

'And it was quite obvious to the farmers, who were treated with utter contempt and disdain, as if they were little kindergarten children that needed to understand. Many, many farmers that I know are highly intelligent, highly sophisticated beings. But they were just treated like kindergarten children. "Swallow this. Here's the Bible, read it in 24 hours, and then follow it and take your lot and be done with it." That was the attitude and it just got farmers really, really angry.'

Farmers were invited to attend an October 2010 MDBA consultation meeting in Deniliquin.

'I distinctly recall going to that meeting,' Robyn said. 'The Authority was taking questions, but there would only be questions for a limited time, and they wouldn't take questions from the farmers in the audience, who were really heated up. So I went along in my business suit, my Giorgio Armani business suit, looking, you know like "Miss Piggy officialdom". Sat up the front and put my hand up. Stood up and then just let them have it.

'"How dare you do this to farmers," I said. "How dare you produce this document and give them 24 hours to respond to it and not produce the other volumes and not allow them to even question the so-called scientific evidence. And that is denial of natural justice." Well of course it brought the house down.

'I just said, "Know where you stand, and you stand in the law."'

Two days after the Deniliquin session, there was another MDBA consultation meeting, this time in Griffith. The views of farmers had only intensified, and the Authority people were warned to expect a tough crowd.

Businessman Paul Pierotti has lived in Griffith for 30 years—almost long enough, he says, to be considered a local. In 2010, he was a principal organiser of a large gathering of people from around the district who were eager to express their views on the Plan. Paul helped arrange for a suitably large venue to accommodate the 13,000- to 15,000-strong audience.

'I was the person to warn the MDBA and the federal police,' he said, 'that if they were to do what they did in Deniliquin, where they locked the hall to a maximum of 200 people when we were expecting some 15,000, that I would not be accountable for their safety.

'I made it very clear. In fact, I'd go so far to say that if we hadn't strongly intervened, I'm not sure that any of them, you know...There was a convoy of about 30 MDBA people, plus the federal government contingent, which was probably five of them. I don't think any of them would have stayed safe on that occasion after what they did at Deni.

'I said they were walking into a minefield. They didn't believe me—until much later, and then eventually they implemented the strategies that I suggested for their own safety. So that included opening up the hall's sides to allow up to 3,000 inside, and putting out speakers and large screens in the carpark and outside to allow up to another 10,000. And then having police brought in from all the regional areas for traffic control and car parking.

'The "presentations"—I would call them that, instead of "consultation"—by the Basin Authority, and all the rest of the garbage that they paddled out at that time—they stuffed around with the terminology to call that release on that occasion the draft guide to the Basin Plan.

'In the first two minutes, Mike Taylor from the MDBA said 7,600 gigalitres would be stripped from irrigators, but this would only equal 400 or so job losses for the entire Basin—a region with over 2.2 million people. Just beyond ridiculous. That day, every national paper, including the *Financial Review*, was showing the complete collapse of Griffith real estate prices and a forecast for the demolition of Griffith's economy, permanently. There were feature articles in every paper.

'So I brought them along on the day. I was, I think, the second speaker when they allowed the public to speak and I brought these to their attention, which they denied at the time, and I said, "Well, you know, we've seen your opinion. We read all about your opinion.

And we believe it to be absolutely, outrageously wrong. For you to keep peddling that in our community is a disgrace."

'The morning session went for three-and-a-half hours. In the middle of that, about 150 people had snuck into the back storage area of the Yoogali Club and collected boxes and boxes of the Basin Plan book. They took them out to the soccer club oval carpark. Then they locked the carpark gate and poured petrol over all the reports and burned them in a massive heap. They alerted all the media, who were already in town en masse.

'The fire was black and stunk of petrol and chemicals, but it was far enough away from where all the crowd was, out the front of the hall. And the people inside weren't aware of it at all, but I was made aware of it by a call from Ray Hadley of 2GB. I was doing radio interviews at the time so it was actually Hadley who alerted me to the fire. I went out and told the private security guards, which the Chamber of Commerce had actually recommended.

'The presentation itself was very predictable and very staged. The real take-home was the reaction. It was the emotion from the speakers. We weren't very happy about all this stuff at the time, and we couldn't believe how naive and stupid the government was. They seemed not to care, and they seemed to not be aware of how pissed off and how potentially volatile the situation was. We were extremely frustrated.'

A theatrical touch added to the dystopian scene, as Paul explained.

'I had hired a group of backpackers to wear animal suits out the front. We were making the comment that you were shutting down food and fibre in Australia. And they were handing out flyers that reflected that—flyers with true data on what the real socioeconomic impacts would be, which up to now had been denied. We'd always been amused by the MDBA director of socioeconomics, whose surname was "towns-end", which we thought was pretty spot on.

'And all of their socioeconomic data has been proven—as we've always said—to be absolute bullshit at every level. They take into account nothing to do with the third-party and flow-on effects. We've always said that the people most affected are not farmers, but people in the community who don't trade or sell water. It's the industries, and the tipping point of those industries where they collapse, so the actual, true economic impact is happening right now.

'We are already estimating it at over $2 billion per annum and this is without all the rest of the garbage that's going on. So it's far worse than that. The latest Sefton report on socioeconomic impacts is the closest thing to actually prove what we've been saying for the last 15 years, but you know, I think the real data is in the numbers and that's already here. And it's only gonna get worse.

'So these "animals" that I had at the front of the hall were wandering about and handing out flyers. Then this guy shows up. A southern Italian guy who is well renowned, well connected. And this is not a movie. You don't want to muck around with these guys.

'So the guy pulls the fabric head off one of the pantomime horse suits and takes it inside the hall. Wearing his checked farmer's outfit, he storms through the crowd. The state member for Murrumbidgee at the time, Education Minister Adrian Piccoli, tries to grab him, but the guy wrestles Piccoli to the floor and then keeps barging forward. When he reaches the stage he chucks the head at Mike Taylor, who catches it and looks at it, you know, face on. Then the Italian guy shouts at him, "That'll be in your bed, you fucker."

'And then right during this session, the government in Canberra announced there would be an inquiry on the Basin Plan, headed up by bloody Tony Windsor of all friggin people. We had to take an iPad up to Mike Taylor on stage and say, "Sorry mate, you've been shut down." All of the Basin Authority people were absolutely devastated and shocked, and they just walked off stage and didn't really know what to do. We quickly herded them off into their cars and got them out of there because I didn't think it was a good idea for them to stick around.'

CHAPTER 28

KILLED

'Investors see water as just another asset category,' Ben Dal Broi said, 'and treat it as such, when it is in fact the lifeblood of rural business, communities and the environment.'

The error at the heart of our water market has imposed a large environmental cost. Counter to the intentions of the National Water Initiative, there are unmeasured diversions and environmental impacts: environmental damage is not reflected in water prices. The implacable westward movement of water rights has exacerbated the intensive use of our natural river system as a water transport system. Too much water is moving, all year round, through the waterways.

According to Laurie Beer, floodplain harvesting in the northern Basin was 'robbing some of the fresh water that should be going down to the lower end of the system, and they're trying to make it up from the Murray system, which is causing a lot of environmental damage;

bank erosion through the Barmah Choke and all those areas. There's an absolute ecological disaster happening there. And the government won't acknowledge this but the locals know. Trees have fallen into the river all over the place because the river's running too high for too long through the summer.

'It's causing bank erosion. Banks get soft, the big trees just fall in, and then that is further limiting the amount of water they can get through it. The old river channel is just filling up with silt and trees.'

'We've dug deeper,' Shelley Scoullar said, 'and uncovered lots of flaws in the modelling of the Basin Plan. That just doesn't sit right with me and what this country's about. So it made me want to go even harder. We actually had a very deliberate campaign of wanting to meet with Malcolm Turnbull when he was prime minister.

'In the end we got our meeting. It was with Scott Morrison, last year [2019], on the day after budget day. We had [federal agriculture and water minister David] Littleproud in there with us, too, and we basically said, "This water can't fit through the system, because of the Barmah Choke. You're trying to deliver large volumes downstream and it can't fit. You're wasting so much water," and he [Morrison] looked at Littleproud at some stage and it felt like, "Why haven't you told me?" Littleproud was trying to justify the situation.'

'He was like a cat on a hot tin roof,' Lachlan Marshall said. 'He was dancing like you like wouldn't believe.'

Shelley continued: 'So in the end the prime minister said something along the lines of, "It is going to take me a while to fix this. I can't fix it overnight. It's going to take a long time to get all the chess pieces in the right spot." So we just had to come away knowing we'd given it our best. We tried to dumb it down the most simple way. The Basin Plan is based on politics, not the best policy. I think we got that message through. I don't know if it made a difference.'

—

When we spoke to David Crew of the Yarkuwa Indigenous Knowledge Centre, he told us how the water had been so high in the Werai Forest that his community had to exhume graves there. He feared the water was eroding the ground and was going to disturb the bodies.

'The result of what was happening,' he said, 'was that a burial site had to be moved because of the water coming through and it was impacting on traditional burial grounds. So those burial sites had to be moved. That was quite a big emotional thing for people. But it's the complexity of a regulated river system where we see the water being kept very high. It's all operational water going through to South Australia, and the rivers don't dry out.

'The natural landscape is one [in which] the rivers dried out and so the banks solidified by being dried out. Now, there's so much water going through, they've just become basically river channels. We have deaths in Deniliquin where people think it's just a river but the currents pushing the water through can pull people under. People have forgotten what the river actually looked like. And we've got scar trees that can fall into the river or climbing trees have fallen into the river. So places that are important and tell a good story are being lost. And then the issue around the protection of burial sites. It sort of scales up. You're talking about the remains where people have been rested and that really upsets a lot of people.'

People in the Basin also expressed concerns that environmental flows were mismanaged.

'I'm of the age,' one farmer said in 2007, 'where I'm a bit more open-minded and understand a little bit more about keeping water for the environment. But we've got a lake behind our properties. In a dry year they put water in it, but it might only last eight weeks. The birds go there and nest, but the water evaporates so quickly. It's like giving someone a house and then burning it down on them two months later with all their stuff in it. I don't think there is much practicality in their decisions a lot of the time. Probably does the environment more harm than good. I'm not going to jump on the bandwagon and say there shouldn't be environmental flows; it just needs to be managed better.'

Unnatural water flows—such as unseasonal cold water flows, and excessive water in summer—account for much of the environmental harm in the river system. Flow management requires a deft touch. Water released at the top of the catchment takes more than a month to reach the coast in South Australia. Managed releases of water from major dams can cause 'cold water pollution'. Fish and other aquatic biota have also died from deoxygenated flows, from algal blooms, and from a shortage of water altogether.

According to the investigation of the causes of mass fish kills in the Menindee region over the summer of 2018–19, there was not enough water in the river system 'to avoid catastrophic decline of condition through dry periods. This is despite a substantial body of scientific research that points to the need for appropriate flow regimes. Similarly, engagement with local residents, Indigenous and non-Indigenous, has been cursory at best, resulting in insufficient use of their knowledge and engagement around how the system is best managed.'

Apart from exacerbating water losses, and causing perverse and damaging water flows, the market design has made environmental flows much more expensive. Thanks to the high water price, taxpayers have had to pay through the nose to return water to the river system. And despite this high cost, a bombshell study published in 2020 showed that most of the environmental water had been stolen or misdirected, or had evaporated, before reaching the intended places. A large share of the steep investment in environmental water was money down the drain.

CHAPTER 29
TURNING THE TIDE

'I don't think you can turn the clock back,' Desiree Button said. 'But I would like to see some more regulations on where water can and can't go. I would like to think that if they [traders] are going to buy it, they have to be able to use it somewhere.'

'There has been a succession of different people appointed to the Murray–Darling Basin Authority,' Robyn Wheeler said, 'and they just try and shut the farmers up. And they won't be shut up. That's the mistake the government has made. Treating them as stupid. And they're not stupid.'

'I struggle to have any optimism for this mess to be fixed,' Ben Dal Broi said. 'Why would it when there is no incentive for the pollies to do so? Most of the irrigation areas are in safe seats, and family farms are not big political donors. Plus, making any meaningful legislative changes to the Water Act is a Pandora's box of inter-jurisdictional argy-bargy that no political party has the appetite to open up. Only when it becomes a crisis will anything be done about it, and it's got to be a particularly massive one because things have been pretty bad for a long time.'

In our water journey so far, we'd pulled off a national first. No one had put all the puzzle pieces together to arrive at the central mistake in laissez faire water trading. But we certainly weren't the only people to get close to understanding the problems. Day by day, people in the Basin saw the market anomalies and lived with the impacts. Outside the Basin, our work coincided with an emerging realisation in political and regulatory circles that something had gone terribly wrong—that the water market we'd ended up with was a long way from the first vision of the National Water Initiative. In official and regulatory circles, the tide was turning.

In 2020, after seeing emphatic social, economic and environmental evidence that the process of buying environmental water from farmers had failed, all the Basin jurisdictions agreed there would be no more water buybacks. (How long that consensus lasts is anyone's guess.) Basin governments and regulators also committed to tightening disclosure rules for water ownership; improving standards of conduct among water brokers; and otherwise enhancing water market integrity and governance.

'It is a commodity,' interim Basin inspector-general Mick Keelty said of water in 2020, 'and yet it doesn't have the same governance and due diligence around it as other commodities like gold or minerals, and I think that's what's of concern to people.'

A striking feature of the ACCC and PC reports on water trading—those from between 2004 and 2018—is how they emphasised benefits over costs. The consideration of benefits came first and was discussed more extensively than the potential costs, which nevertheless were diverse and very considerable. Today, when examined with the benefit of hindsight, the ACCC and PC documents read as though they were advocating a foregone conclusion. Rules about inter-valley trade, for example, were characterised as 'restrictions' rather than safeguards. The rules were presented as assaults on efficiency and liberty, rather than protections for farmers and communities. The documents read like sales brochures for a free market.

The ACCC and PC used lazy analysis and lazy language. A good example is the ACCC's first *Water Monitoring Report*, from 2009–10: 'Barriers to trade imposed by governments, such as Victoria's 4 per cent annual limit on water access entitlement trade, still distort market outcomes.' The phrase 'distort market outcomes' is a universal, neoclassical crutch, used to avoid a genuine accounting of costs and benefits.

In light of that history, the ACCC's 2020 water inquiry report was an extraordinary reversal. In the interim report, the ACCC acknowledged the market was far from transparent and that it lacked integrity. The ACCC stated, for example: 'While water market transparency is one of the objectives of the National Water Initiative, its implementation has been haphazard due to Australia's multi-jurisdictional approach to water reform and the disjointed way that trade-related services [such as water brokers] have evolved.'

The ACCC highlighted a variety of market gaps and information problems: 'There are currently many disconnects which impede the free flow of core market data from its generation source through to end users who use this data to inform their own trading decisions or provide advisory services to others. This has contributed to water market participants' lack of confidence in water markets and variation in what "transparency" means to different stakeholders.'

It is extraordinary that these findings could be made in 2020, more than a decade after the market commenced. The change of thinking at the ACCC was driven in part by a change in personnel (such as the appointment of Mick Keogh as ACCC commissioner in June 2018) and in part by the Hayne banking royal commission, which green-lit a more activist role for regulators, caused more scepticism of bankers, and shrank the influence of 'ultra-dry' economists in policy and regulatory circles. Post-Hayne, the ACCC has become decidedly less Friedmanite in its views on the water market.

If the ACCC were a politician, or in fact any other type of natural person, it would be forced to explain this gigantic flip-flop. As a public body, however, it can hide in the bureaucratic grey zone it occupies between the parliament, government departments and the water market, and it can make the most of the minimal scrutiny that regulators and quasi-regulators attract in Australia. But, notwithstanding the pirouette, the ACCC (and the PC) must bear some of the responsibility for Australia's epic water fail.

In 2020 we arranged to meet Malcolm Turnbull, former prime minister and former federal minister for the environment and water. In January 2007, Turnbull had put $10 billion on the table to support the National Water Initiative. 'The 10-point plan,' the *Sydney Morning Herald* reported at the time, 'includes the biggest modernisation of irrigation infrastructure in Australia's history.' When we spoke to Turnbull, he was in a relaxed and reflective mood—on track to becoming the wise and generous figure that many progressives wished he'd been as PM.

'I spent a bit of time,' he said, 'talking to Labor's Tony Burke about water, he's a completely non-ideological person. He's a classic product of the New South Wales Labor Party right. As Neville Wran used to say, "a Bible in one hand and a dagger in the other". I think that's a bit harsh on Tony, but he is essentially a pragmatist—and they won back a bit of confidence [with farmers]. But water policy is hard. There's no other way of describing it. It is bloody hard. We've got less water and the claims on it are as great if not greater than they've ever been.

'People in Treasury would say, "Oh, you know, we've got to have a water policy managed by economics not politics." I'd say, "Well that's great, but you have to eliminate all human beings if you want to get the politics out of it."

'Depending on how farmers answer these questions, maybe they should focus on products like wheat or, dare I say it, cotton—which everyone cringes about. Believe me, I'm not a defender of those crazy diversions, but one of the advantages of cotton is that if there is not water, you leave the seed in the shed. But if there is no water and you're a dairy farmer, what do you do with your cows? What do you do with your fruit trees? It's complex. Water policy is hard.'

CHAPTER 30
UNSCRAMBLING THE EGG

The year 1989 was the start of organised water trading in Australia. It was also the end of communism in Hungary. Veronika Nemes was a university student in Budapest at the time. She saw first-hand the city's rapid switch from central-planning grey to free-wheeling technicolour. Las Vegas Casino, Budapest, was a symbol and an example of the hyper-capitalism that followed communism.

In 1992, Sylvester Stallone was at the height of his fame, and the casino booked him for its grand opening. American *Playboy* was another symbol of change. The magazine had recently appeared on Budapest newsstands, and to promote the casino's opening, the magazine included vouchers that could be cashed in for gambling chips.

The casino was as Vegas as the real thing, but Veronika noticed the promoters had made a mistake. They'd mispriced the Stallone vouchers, which were too generous. The fine print, moreover, placed no restrictions on their use. If someone bought a lot of the magazines, which had a cover price of around a dollar, they could make a killing by collecting the vouchers and cashing them in at the casino. So that is what Veronika did. Going from newsstand to newsstand with a large plastic bin, she engaged in some unregulated arbitrage.

Buying hundreds of *Playboy* magazines, she cut out the coupons, re-sold the magazines (for only a modest loss) and took the paper loot to the casino (for a big gain). Eventually the casino wised up, and the owners limited Veronika to redeeming her remaining Italian Stallion vouchers for food at the casino's restaurants. Thus well funded and well fed, Veronika moved to Australia, where she completed her PhD in economics at the University of Melbourne, and became a recognised expert in markets. One of the few people in Australia in the 2000s who really understood market design, Veronika joined the Victorian government's specialist markets unit, led by Gary Stoneham.

With Professor Peter Bardsley of the University of Melbourne, Gary had assembled the market-design A-team. They drew concepts from the burgeoning international fields of auction theory and game theory: ways to combat collusion, improve information, and set prices for assets that have multiple dimensions of value, such as environmental assets.

For the use of markets in public policy, those tools were invaluable. Early experiments in this field were remarkable for how they failed, mainly because participants came up with clever ways to 'beat the house' and game the system. Examples included aggressive trading in unregulated energy markets, and collusion in telecommunications spectrum auctions.

The A-team experimented with trading environments and

governance structures that would prevent collusion and other forms of monopolistic conduct. They designed mechanisms that would make full use of market information, and elicit truthful disclosures from market participants. And they used ingenious fractal maths to measure biodiversity and other dimensions of environmental value.

Armed with these tools and this knowledge, the team worked on policy assignments in energy, environmental services and procurement markets. One of the best examples of the A-team's work was the BushTender. Constructed around an innovative computer interface and algorithm, this was a market-based mechanism for setting public subsidies to protect remnant native vegetation on private land in rural and regional Victoria.

Because of the history of market-based policy tools, the BushTender designers had to take great care to avoid the pitfalls that were prominent in early market design experiments. Trialled from 2001 to 2003 and fully implemented in 2006, the BushTender melded auction theory, an electronic bidding platform and old-fashioned public administration.

A very noticeable feature of the auction rounds was the strong integrity framework: there was a methodical opening of bids, as well as scrutiny of all bids by a probity auditor, who was present at all times to confirm the rules were followed. Inspectors were sent to check that farmers' proposals were valid and to score them: how much land and biodiversity would be protected, and of what quality, as assessed against ecological criteria?

The inspectors themselves were subject to probity checks. Probity auditors had to grapple with curly ethical questions. On one inspection visit to a farm in central Victoria, for example, the farmer offered the inspector an opportunity to buy chickens. The inspector had to seek a ruling from the BushTender probity auditor as to whether the inspector could buy the live birds from the farmer. (The answer was no.)

A crucial feature of the auction system was how it treated time. Because the BushTender was held in rounds, the specific order in which bids were submitted and received didn't matter. There was no race between farmers. The only thing they raced against was the clock: there was a deadline for submissions, which the probity auditors strictly enforced.

Equally as important, there was no scope for predatory behaviour or for gaming the system. Thus designed, the BushTender won the trust and the confidence of participants, and it became a powerful mechanism for protecting native vegetation. Ideas from the BushTender spread to other settings further afield, including river health auctions, capital works tenders, and efforts to save wild tigers in South-East Asia—by changing the incentives of poachers.

In our research on the Murray–Darling Basin water market, we'd unpicked the problem, and we knew policymakers and other influential voices were getting close to understanding it, too. But the obvious question was: what happens next? The ACCC, PC, farmers' federations, irrigator groups, lobby groups, politicians, journalists and others had all posed the same question. Where to next for Australian water?

On our water journey, we'd seen and learnt a lot: about how the market had failed, how the failure had affected farmers, and what a future water world should look like. Based on the experience of people in the Basin, any future water model would need to solve at least four core problems.

First, it must reduce the tendency towards corruption. A future model needs strict rules for water accounting and integrity in the allocation and use of water. It has to be fair and transparent.

Second, a new water model needs to address the stark power imbalance between groups of participants, and in particular between farmers and professional traders. A large part of improving the power balance involves improving information. Another big part involves slowing everything down. Much of the traders' clout comes from their speed, especially their ability to be first in the market and first in the queue with inter-valley transfers.

In the interests of fairness and integrity, the future model should not be based on real-time trading. As the agricultural investor goFARM noted, the 'fastest fingers' element in the market 'favours a few and disadvantages many'. We believe there are viable ways to make speed much less important.

Third, the water model needs to reflect the multiple values of water. For people in the Basin, and for Australians overall, that water has significant social, cultural and environmental values, as well as economic ones. Water allocation decisions need to reflect considerations and criteria such as economic growth, productivity, biodiversity, food security, streamflow management and fairness.

And finally, a new water model needs to recognise the tendency for financial returns to dominate economic ones in the Basin. Water for irrigation should be used in irrigation, and it should be used where the highest national value can be achieved—not just where an astute trader can reap the highest private financial return. Solving this problem is not as simple as allowing only farmers to trade water. As we have shown, traders can pretend to be irrigators by using versions of the Holy Grail strategy.

In 2020, former prime minister Malcolm Turnbull commented on our then tentative proposal to redefine water rights for irrigators: 'I understand what you are proposing,' he said. 'Keeping water rights separate from land, but redefining the entitlement as "water in use"—in other words for irrigators only. The thing you want to be careful with is this: limiting how people can use their asset will tend to diminish the value of the asset. So you might have some pushback from irrigators there. It's a bit like, you know, farmers saying "I don't like all these foreigners buying land in my district." Well, that honestly depends on whether they're a buyer or seller. The minute they want to sell, they don't want any buyers to be disqualified.'

To solve these four problems, wholesale reform is needed. Australia needs to adopt a different mechanism for allocating water across the Basin. That mechanism needs to reflect the multiple values of water, including Indigenous values.

One potential model, based on the BushTender, would feature a single trading platform, bilateral tenders and a central body that would apply multiple criteria to reflect the multiple values of Basin water. (In a bilateral tender model, farmers would submit bids and offers, and the central market body would rank and match them.)

Such a managed market would internalise the complexity, and treat participants equally, including by removing the flash-boys element of 'first in, best dressed'. (For example, the bids and offers could be 'sealed' and submitted before a daily deadline.) It would also capture the currently unpriced impacts of water allocation decisions, such as water losses, social harm and environmental damage, and it would take account of capital market biases, such as anomalies in taxation and differences in access to capital markets.

That kind of model could continue to function within the framework of capped diversion and allocating water for the environment. In addition, it would capture a potential synergy that is lost in the current market design: the synergy between environmental flows and the movement of water between regions for irrigation.

The multi-criteria assessment of bids and asks would also internalise intra-Basin 'exchange rates' that take into account differences in costs and values across different regions and valleys.

Such a model would restore the purpose of the water market as a water allocation mechanism, not a financial market. But we think there is a strong case for even more significant reform. Given the extent of the market failure and its impacts, and given the importance of water now and even more so in the future, we see a case for stepping away from the use of markets to allocate Basin water.

Alfred Deakin sketched an alternative to market allocation. In a Deakinite model, a citizens assembly—of people drawn from across society and across the Basin, with a strong Indigenous voice—would make the high-level decisions about how to allocate water in the Basin. (As the finders of the Templar Treasure say in the Nicolas Cage and Diane Kruger film *National Treasure*, 'Give it to the people.')

The assembly would apply the social, economic and environmental criteria in their decision-making, and they would make judgments in the context of the circumstances at the time—such as drought, pandemics or other threats to Basin health and risks to food security. Complementary to that high-level allocation of water, there could still be local water trading between farmers and within districts. In a partial return to the 'water for beer' era, farmers would enter formal and informal bilateral agreements with each other, and those agreements would be captured in a new, streamlined system of water accounting. (Farmers could also bid for water in local auctions.) A model like this would reinstate the nexus between water and place.

Fights over water are a defining feature of modern Australia. One cause of our epic water fail was the decision, at the time of Federation, to split water across individual states rather than to properly manage water at the whole-of-Basin level. The rail gauge problem was always going to be a problem in water, too, and it explains why we have a steaming mess of rules and anomalies across jurisdictions. (We are not suggesting the water powers should go to the Commonwealth government, just that there should be suitable inter-governmental arrangements.)

The new water model should be designed and governed through bipartisan agreement, so it can outlast particular governments and partisan trends, just as the Howard-era gun reforms have. Another example to emulate is the bipartisanship at the beginning of the COVID-19 pandemic, epitomised by the National Cabinet, which was much more effective than previous COAG and National Competition Policy forums (not a high bar).

We recommend that a reconvened National Cabinet should oversee the Murray–Darling reforms. Further, a cross-party, cross-jurisdiction standing committee should be formed, reporting directly to the National Cabinet. The committee and the Cabinet should listen to diverse voices including Indigenous and non-Indigenous people from across the Basin, plus other specialists nominated by each party in the senate. (Prime Minister Julia Gillard set a precedent for that in 2010 when she formed the parliamentary committee on emissions trading.)

This diverse and inclusive governance process would complement and support the diverse and inclusive citizens assembly. Under this model, reflecting Deakin's vision, we would decide as a society how best to allocate scarce water in a way that balanced all the imperatives such as efficient agriculture, river health, and recognising the cultural and economic rights of Indigenous people.

—

Indigenous people have long seen water and land as integrated parts of a whole. Indigenous water management, land management and regenerative farming techniques will be crucial if Australia is to live with climate change, droughts and firestorms.

'Jeanette was the first female Aboriginal site officer in New South Wales,' David Crew told us, 'and then she worked in the Lands Department and other agencies, moving here in year 2000 to take up a position as Aboriginal natural resource officer for the Murray region. That gave her the opportunity to be talking with landholders about cultural values on their land as well as understanding the values that exist within the environment, while farmers were also doing their farm work.

'Often water is put in terms of cultural water or environmental water and it has often not been expected that that should make an economic return. So we talk about the concept of cultural economy. And the fact that there is an Aboriginal cultural economy around traditional foods and medicines and plants and if they're not allowed to make an income out of that, because of where the water's come from, that is a contravention of human rights.

'Cultural water is about determining when water needs to be in a particular location for a cultural outcome. That can only be achieved by using the knowledge of the people. Water should not be a competitive commodity. Rather, we accept that water that flows through our region can do multiple things as it comes through. Unfortunately, this has been really problematic with the separation of water from land. And then the commodification of water, which then becomes making water a valuable commodity, and then we get these investors making money out of it. This is where we are now, which is really what stimulated a lot of pain for a lot of people because unless you've got a lot of money you can't buy the water.

'The Murrumbidgee has a cultural water allocation. That allocation generally has been provided to the community around Hay. The biggest issue they had at the time was that while they had the water they didn't have the resources to actually pump the water. They said, "Here is your cultural water but we're not going to help you get the water to where you need it to be."'

Apart from making poor use of Indigenous knowledge about water management, the multibillion-dollar water market is one of several spheres in which Indigenous people have been disadvantaged and dispossessed. Phil Duncan, a Gomeroi man from Moree, recently called for greater Indigenous involvement in the water sector, founded on 'self-determination, self-respect and self-management, particularly in the water arena where we have been somewhat operating in a vacuum and for quite some time'.

Two centuries after the start of European colonisation, there was serious talk of a proper accord between black and white Australia. In the bicentennial year, for example, Bob Hawke visited the Northern Territory and stated that there would finally be a Treaty. Fast forward another three decades and there is no such accord in place. Now is the time to put water on the table at Treaty negotiations and in the conversation about an Indigenous Voice to Parliament and the Uluru Statement from the Heart. Genuinely closing the gap requires genuine reconciliation, and that means Voice, Treaty and Truth—the core elements of the Uluru Statement.

One day soon, we hope, Indigenous water rights and widespread adoption of Indigenous water management practices will be as solid and permanent as the eel dams of Budj Bim and the fish traps of Nyemba.

CHAPTER 31
THE BETRAYAL

As part of our water journey, we saw a lot of creeks, rivers and Basin communities. We swam in the Murray at Echuca and Yarrawonga and Mungabareena; the Murrumbidgee at Wagga Wagga, the Howqua near Jamieson, the Ovens at Bright, the Buckland at Puckapunyal, the Yarra at Warburton, the creek at Yackandandah and the open cut at Allans Flat. We fished for redfin and yellowbelly in the irrigation channels at Tatura, for trout (with a weighted brown nymph) on the Goulburn, and for carp at Mildura.

We rode dirt bikes at Urana, camped with emus and wombats at the edge of the Blowering Dam, explored the floodplains of Numurkah and Wangaratta, and marvelled at the grand junction of the Murray and Darling. When the Hume Dam was very low we found old wagon wheels, dead turtles and half a $2 note, still bankable.

We drove from Bendigo to Mildura via Swan Hill. From Albury to Tamworth via Dubbo. From Mernda to Jamieson via Yea. From Cohuna to Coonabarabran via Parkes. And from Wentworth to Broken Hill via the Menindee Lakes.

Many of the creeks, rivers, farms and communities we visited are at risk thanks to Australia's failure in water policy.

After embarking on our water journey, we were genuinely and continually amazed at what had happened. Australia went hard and early into water markets—much faster and further than other states and countries around the world with similar problems, such as California, China and Chile. Australia pushed water market reform so far, it surprised foreign water-policy people—especially given how important water is in Australia.

Foreign bankers were just as surprised, and they were delighted. For investment banks and hedge funds, the concept of investing in water was irresistible. It allowed them to coat ordinary greed in a green veneer. Governments let the banks and quasi-banks into the water market with the intention that those participants would provide 'liquidity'. Once inside, the traders levied a handsome fee for this service—a fee that was gathered through small and large arbitrage. The liquidity was expensive, and it was of low quality—an example of 'junk liquidity'.

As Nobel Prize winner Joseph Stiglitz pointed out in a 2014 paper, arbitrageurs in financial markets may provide liquidity, but they do not provide genuine demand: they are merely seeking profits by interposing themselves between the actual trading parties—as is the case when a trader buys water then quickly sells it to a ready buyer in the market, after taking a cut.

That kind of liquidity is not available when the market needs it most, such as in a crisis or some other kind of collapse. The high-speed arbitrageurs are simply stealing market rents. And the providers of junk liquidity, Stiglitz argued, can actually reduce genuine liquidity, for instance by damaging confidence in the market. They can also reduce liquidity by holding assets off the market, in pursuit of longer-term trading profits.

Stiglitz was writing about conventional financial markets, but the problem of junk liquidity is even more severe in the water market. The prices of water rights differ markedly over time and between valleys. Due to the high volatility, and the disjointed nature of trading, the water market is in a permanent state of disequilibrium. Junk liquidity in water is even junkier than in the stock market. Fast traders can reap arbitrage profits until the cows come home.

Unlike stocks and bonds, water is an essential input with few if any substitutes. Its value is highly seasonal, and highly contextual. It has a fixed supply, and that supply is dwindling. When policymakers speak of external traders enhancing 'liquidity' and 'efficiency', they are applying ill-fitting constructs and a false comparison between conventional financial assets and financialised water.

Basin governments turned water into a financial asset without paying attention to market conduct, and without applying the market protections and regulatory safeguards that apply in conventional financial markets. The ability of water traders to manipulate the market exacerbated the volatility, accelerated the arbitrage opportunities and perpetuated the disequilibrium.

In addition to applying false analogies, policymakers confused critical concepts. Under the banner of liberalisation and deregulation, governments allowed the gaming of the water market. Manipulating the rules of the game became a routine strategy, and a lucrative one.

To better understand the epic failure of Australia's water

market experiment, we spoke to financial innovation specialist Ian Shepherd. Formerly a member of McKinsey & Company's international banking leadership group, Ian left that famous firm to design a new type of high-integrity financial market. From multiple viewpoints—including in the celebrated court case *Alice v CLS*—Ian saw Wall Street banks and traders up close, and he could see what had happened in Australian water.

'Governments allowed competition *within* the water market,' Ian said, 'but also competition *for* the market. Any market will fail without a strong framework and clear boundaries concerning how participants can trade.'

The water market designers were naive to think Wall Street traders and other external investors would merely be a benign force in the market, gently providing liquidity. The combination of a lax regulatory regime, high-speed trading tools and aggressive tactics was only ever going to lead in one direction. The traders did everything they could to strip profits from the market. But there was nothing surprising, Ian told us, about that fact.

'Of course the traders sought arbitrage opportunities,' he said, 'and played the market. The traders were just doing what they do. Pursuing opportunities to make money. It's second nature.'

The international banks and quasi-banks had entered the Australian water market as a bit of a punt. But around the time of the global financial crisis, the larger water traders realised the market was ripe for a grander strategy. That strategy was never written down, and the traders' backers and employers would never admit that they had adopted it. We don't know for sure who came up with the idea in the beginning, though we could make a good guess.

Through our interviews with brokers, traders, farmers, officials and exchange managers, and by analysing corporate disclosures, regulatory reports, policy reviews and even the underground literature of whistle-blowers and activist short sellers, we've been able to piece together a picture of the strategy and how the traders sought to implement it. The strategy brought together all the elements that were already present in the financialised water market—including bots and market manipulation—and it introduced new ones.

Thanks to the Holy Grail strategy, the market was a bonanza for water traders—and a massive net loss for rural communities. Financial returns from Murray–Darling Basin water trading are consistently higher than those achieved in agriculture and even in the stock market. External water traders have asset-stripped a generation of investment from Australia's foodbowl towns.

As a Basin farmer said: 'We've forgotten the goal of the market—to drive an efficient allocation, not to enrich smart traders.'

In our research, we've traced multiple similarities between unbundled water rights and the sliced-and-diced securitised mortgages. And here is one more mortgage analogy: the Basin farmers have adopted a 'buy and hold' position akin to the old-style commercial bankers who, before securitisation, held mortgages on their balance sheets. In contrast, the active traders and brokers are not interested in 'buy and hold' returns, and many of them viewed the farmers the same way they saw the old-style bankers—as dopes and slowpokes.

We also traced parallels between water trading and the worst excesses of energy trading. In the 2000s, American electricity traders welcomed blackouts. Now, big water traders welcome droughts. And there are few things more reliable in Australia than droughts.

Between 1989 and now, farmers have become more sophisticated in how they trade water, but IT and finance have leapt ahead at a much faster pace. Over the past three decades, there have been recurring FinTech revolutions—in risk measurement, insurance, contingent-state securities, hedging, behavioural economics, swaps, high-speed trading, big data, exotic financial products, and ever more complex types of 'bet' that traders can make. Every year, the gap between farmers and traders has widened. In no sense are farmers catching up.

If we don't fix this imbalance soon, the foodbowl farms will all go. The pursuit of an 'efficient market' (an economic solecism) risks turning the whole of Basin agriculture into industrial farms, and all the independent farmers into employees. Regulators say the market is maturing, but it has been maturing since 1989; and in the meantime a lot of farms are closing and a lot of money is rushing out of the Basin. As every undergraduate economist learns, in the long run, we're all dead.

Australia has failed to deal simultaneously with the tragedy of our climate and the 'tragedy of the commons'—how best to make use of a resource that is both shared and scarce. The Australian water market is the biggest public policy disaster in living memory. It is a 'triple bottom line' disaster: economic, social and environmental.

Failures in water management are happening at the worst possible time. As the world's population surges, and as there is more competition for arable land, the viability of foodbowl regions is increasingly urgent. Climate change is only making water more important and more scarce.

Amid the COVID-19 global health and economic shock, we've seen Australia's vulnerable supply chains exposed. Simultaneously, in the coldish war between the US and China, Australian exports were in the firing line. Food security is a major risk for Australia, and the diversity of our agriculture is crucially important. We can't all live on grapes, olives and almonds.

The story of our water market is a story about the failure of Australia's system of public administration and government. It is an instance of what economic journalist Martin Wolf recently called 'rigged capitalism' and 'rentier capitalism'—phenomena that are degrading public governance and undermining democratic capitalism.

In 1903, Alfred Deakin became the second prime minister of Australia. He saw irrigation as a three-way partnership: it was simultaneously a private, state-authority and community responsibility. For Deakin, the prospect of water becoming a private monopoly was unthinkable. As described by J. M. Powell in his book *Watering the Garden State*, Deakin's report after his California trip included this recommendation:

> To guard against the rise of local monopolies and the ruination of valuable schemes by the activities of disaffected individuals and small minorities, each irrigation area should have a means to set up an organisation and operating structure answerable to the bidding of the majority, and capable of raising funds on the strength of a common security.

How did Australia, collectively, make the social decision to allow the water market to go ahead largely without fetters? And how, in particular, did we forget Deakin's advice about guarding against the emergence of water monopolists?

As often as not, when making decisions about water, successive governments listened to the louder voices of glamorous 'experts' from the US, the UK and New Zealand, and voices from international accounting firms and investment banks. But as far back as the 1990s, a lot of the key water market design pioneers—including some of the most highly paid academics and consultants and parachuted-in officials—presented themselves as economists and market design experts when in fact they were not.

Market design is a newish field. At the beginning of the water market, there were no market design guilds or accreditors, and the price of entry was low, just as it was for water brokers. Advising on water trading became a picnic for regulators, consultants and other experts, none of whom protected Australia from the failures to come. As well as pretending to have relevant qualifications, they pretended they had the necessary tools, including ones that would ensure market integrity and protect smaller farmers and rural communities.

Top-level regulators and agencies such as the Murray–Darling Basin Authority, the Productivity Commission and the ACCC assembled and published report after report. Between 1989 and 2020, more than 300 major water sector reports were produced in Australia. In retrospect, many of the volumes read like case studies in how not to do policy and research. Brimming with fuzzy logic, linguistic ambiguity, categorical confusion, statistical insignificance, motherhood statements, statements of the obvious and statements of the obviously false, they are now useful only for recycling and for counter-examples in teaching.

Rural communities have also found another use for them, as kindling, but we don't recommend using the reports that way. There are too many chemicals in the paper, and the reports burn too wildly. All the reports and reviews and studies did, however, serve a useful anthropological purpose. In the same way that traders' megafarm partnerships were a fig leaf, the official reports provided political and bureaucratic cover for a giant stuff-up. A once-in-a-generation scandal. A government-authorised mega-theft.

The water market looks arcane and mysterious, but it is not a magic trick. Over the past three decades, water policy in Australia has been a shell-game fraud with three cups: the legitimate veneer with 'liquidity' traders and agribusinesses; short-term arbitrage trading and market manipulation; and finally, and most lucrative of all, cornering the market through the long-term Holy Grail strategy. Today's water market consists of the respectable-looking front, the short con and the long con.

The big drought that struck south-eastern Australia in the early 1880s led the colonial governments of New South Wales, South Australia and Victoria to establish royal commissions. Two decades later, in 1902, the first interstate water royal commission was assembled. Over the next 100 years, a succession of major inquiries into the Murray–Darling Basin would contribute to the ever-growing library of water reports. The most recent example, a 746-page brick from South Australia's Murray–Darling Basin Royal Commission, was published in January 2019.

That same year, we said in the *Mandarin* that Australia didn't need another water royal commission. Today, we still believe we don't need such an inquiry to find out what went wrong in the water market, and how our political and policy system betrayed farmers and rural communities. The mistakes and missteps are clear.

We've suggested that traders and megafarms may be inclined to destroy evidence of their coordinated activities. Certainly monopolistic conduct is an offence in Australia. But one of the most disconcerting findings from our research is that many of the market phenomena we've documented are permitted at least to some extent by the water market rules. Our representatives collectively allowed the epic fail to happen; they let the wolves and barons inside to play with our most precious natural resource. Any attempt to prosecute the traders for those activities would very likely fall short.

But we believe a different kind of inquiry is now needed: one with open hearings and strong powers to 'follow the money', to fully understand the extent and impact of the policy failure. We have made the discoveries in this book just by talking to people, listening, and joining the dots. We did not have any formal powers to discover or seize commercial documents or to delve into the management accounts of major traders. A suitably constituted body with those powers needs to follow the trail we have described, and to do what can be done to unwind the catastrophic errors.

In following the money, the inquirers could use some of the water traders' own data-heavy, IT-savvy methods, such as by using big-data techniques to sort through all the historical trades and the metadata held by exchanges and irrigation bodies. That way, we would have a clearer picture of all the different types of transactions; the relationships and arrangements between traders and agribusinesses; where all the money went; and to whose benefit.

There are strange, recurring patterns in water prices, possibly evidence of fractal processes. We've not confirmed that, but we did find fractals in the water market. The river system is fractal in its structure, and so is the market—in its failure. The market is broken in ways that replicate at multiple scales. Daily, monthly, yearly arbitrage. Small-scale and large-scale market manipulation. Small and large-scale categorical confusion. Small and large-scale theft.

Australia's epic water fail is a scandal so outrageous it can be smelt and felt and tasted. In this book, we've outlined a model for urgent, wholesale reform of river management, one that is much more BushTender and much less Budapest Casino. How Basin governments respond to our findings will be a test of our system of democracy. Will that system continue to treat the Basin's leaders as dopes and troublemakers? Or will Australia finally hear their desperation? And will we finally hearken to their deep reserves of hard-won wisdom? To change the flow of Murray–Darling Basin water policy, the first step is easy. All we have to do is listen.

EPILOGUE

David Crew: 'As bad as things are,' David Crew said, 'I see a window of opportunity. Understanding that we can share values. We're seeing a window of opportunity here at our high school. Next year, all of the year sevens will be doing a language and culture stream. Every year seven for three periods a week, the whole year, will be doing something relating to local language and culture. And then year eight are doing what's called the sustainability stream. They're working in our local wetlands and in our lagoons in Deniliquin. It's a culmination of ten years of work and it's actually changed the amenity of the local town.'

Damein Bell: In 2010, a 70-metre-long weir wall (with a 'fish ladder' for eels and fish) was completed to fill the lake of Budj Bim. 'The government was really good working with us,' Damein Bell said, 'in how we wanted things done in regards to the hydrology, the digital landscaping, but each time you sort of had this man come down and he said, "Well, how are going to work out your water allocations?" So we would say, "Well, we will get it from our ancestors. Where do you get yours from?" And he sort of scuttled away.'

The last time the lake had been full was during the 1946 floods. Within two weeks of the weir being completed, rain filled the lake in a single weekend. 'It was amazing to see the water snake across the dry lake-bed to hook up with traditional infrastructure on the other side...The restoration has worked wonderfully. Even in the middle of summer, when it got to 37 degrees, the lake still had a good body of water in it and there were still birds there. The mob have been out there fishing.'

Ben Dal Broi: Ben continues to advocate for sensible water market reform, including through the Benerembah Warrawidgee Water Users' Association. 'Governments are hostile to the idea of a royal commission,' Ben said. 'I'm not in love with the idea as it is expensive and drawn out, but I can understand why it appeals to many people. Royal commissions are usually demanded by people desperate to be heard pointing out a major systemic failing that the current authorities either can't or don't want to fix. Think "The emperor has no clothes".

'When I asked federal politicians for their views on a royal commission, one said, "Be careful what you wish for," in that farmers could potentially be worse off. But from my perspective it's hard to see how irrigators could be worse off.

'Sadly we're the victims of a massive experiment gone wrong. To fix one set of problems, governments have created bigger ones by rushing the forced implementation of a deregulated market system.

However, the Basin isn't a perfect market. There have been a small number of massive winners, and many losers, including the communities and the environment. The irony is that they are seeking to re-regulate the market. Look, I support necessary reforms, but to properly fix something this big and complex needs to be done with support of good science and common sense, and done cautiously, over decades.'

Shelley Scoullar: Cereal company Kellogg's featured Shelley in the 2018 'ode to a farmer' series to celebrate 100 years of using Australian grains. The company sent an acclaimed virtual reality artist to render Shelley's farm digitally. Subsequently published, the rendering was a wonderful celebration of Shelley's achievements and her love of farming. But in the new world of Basin agriculture, even dedicated and passionate farmers were pushing uphill. 'In 2019,' Shelley said, 'we had to make some really tough decisions. For me, growing rice was what I loved and was passionate about. But I felt we weren't going to be able do that year in, year out. We made the tough decision to sell the farm.'

Shelley has mixed views about the future. 'We have great opportunities,' she said, 'for farms, towns, the environment, our ecology, Indigenous reconciliation—but we have to make it happen or we're doomed.'

Long-Term Capital Management (LTCM): After a run of annual returns of 30 per cent or more—due to extreme risk-taking and super-complex arbitrage trading, much of it via the over-the-counter market—LTCM's returns fell into deep negative territory. It turns out there were fundamental errors in the maths that the arbitrageurs were using. Very awkward for the founders, including two Nobel

Prize–winning economists. LTCM was bailed out, and it earned a new name, 'short-term capital mismanagement'.

Enron: The collapse of Enron in 2001 led to a series of prosecutions and subsequent appeals. Many of the original allegations made against Enron's leaders and the auditor Arthur Andersen turned out to be false. Knowing this, the US government, its regulators and the legal system were eager for the matter to go away. The auditors were ultimately cleared, as were other principal players in the affair. Not only were Enron's tactics largely lawful, they were very similar to what Wall Street banks did every day of the week.

APPENDIX: TIMELINE

c. 100,000 BCE to today: The oldest fossil of the modern platypus dates back to around 100,000 years ago. The *Ornithorhynchus anatinus*, or the duck-billed platypus, is a unique egg-laying semi-aquatic mammal. It eats annelid worms, insect larvae, freshwater shrimp and yabbies. Unlike for the kangaroo, a pouch would be a drag in the water, so the platypus hides with its young in burrows. The mostly nocturnal monotreme feeds her babies with milk, senses prey through electrolocation, and conceals a sting on her hindfoot.

c. 70,000 BCE to today: For 70,000 years or more, people have lived on the Australian continent, building a rich tradition of art and culture and assembling a deep body of knowledge—which is science and engineering as well as culture and tradition—about water, the land and natural resource management.

In recent times, Traditional Owner organisations have noted that treating water as a tradable property right is at odds with many Indigenous beliefs. According to the Murray Lower Darling Rivers Aboriginal Nations submission to the ACCC's recent Murray–Darling Basin inquiry: 'For many First Nations peoples, the separation of water from land, the formulation of water "products" as commodities that can be held and traded for private profit and the disembodiment of water from its sacred and spiritual contexts are fundamentally at odds with deeply enshrined water values and custodial responsibilities.'

Jeanette Crew, Mutthi Mutthi Elder, wrote in relation to the Edward/Kolety and Wakool floodplains and forests: 'These forests were our economic base for thousands of years and now provide no economic return for my people, while at the same time making many non-Aboriginal people wealthy. My people's spiritual and religious connection to country are directly linked to, and cannot be separated from, the environment.'

Some Indigenous groups (such as the Wadi Wadi Nation on the Murray) adopted the platypus as a totemic creature. Platypuses feature prominently in the Dreamtime stories that have long been shared on the banks of the Darling River.

1800 BCE–1500 BCE: Instances of water theft can be traced back at least as far as ancient Mesopotamia. Around 1800 BCE, laws were passed there to protect water flows through irrigation channels. Similar Hittite laws from 1650 BCE to 1500 BCE stipulated that people who diverted water from an irrigation ditch had to pay compensation.

c.343 BCE: The Greek philosopher Plato wrote about water theft: 'Water above all else in a garden is nourishing; but it is easy to spoil.

For while soil and sun and wind, which jointly with water nourish growing plants, are not easy to spoil by means of sorcery or diverting or theft, all these things may happen to water; hence it requires the assistance of law.'

58 BCE–50 BCE: Water affected the rise and fall of empires. Rome's power and prosperity were underpinned by ingenious systems of aqueducts, drains and sewers. Julius Caesar's victories in the Gallic Wars depended in part on his use of engineering and technology, which enabled him to cross channels and rivers. Caesar's bridges across the Rhine river were an engineering and military masterpiece—allowing his legionnaires to dominate the landscape.

332 CE: Three centuries later, a farmer called Sakaon noticed his upstream neighbours were illegally diverting water in Egypt's Fayyum Oasis.

c.1200–1400: In the Middle Ages, churches kept holy water under lock and key because, paradoxically, it was an ingredient in witchery and black magic.

1850s: First pastoral leases and irrigation schemes on the Murray. Paddle steamers on the rivers Murray and Darling.

1851: News of the discovery of Australian gold circles the world. The booming Port Phillip district becomes a separate colony. Rumour that when Port Phillip was to become a separate colony, London officials chose the Murray River as the northern border because the alternative, the Murrumbidgee, was too much of a mouthful, and didn't really sound like a proper word.

1883: Construction on the Torrumbarry system—beginning of diverting water from the Murray to Victorian irrigation. Conference for Australian Federation held in Corowa, NSW.

1887: Chaffey Bros establish irrigation settlements at Renmark in SA and Mildura in Victoria. Conflict between the colonies over threat to navigation in SA caused by upstream water diversion.

1895–1902: The Federation Drought.

1897: The only 'lock and weir' on the Darling River is built (at Bourke, NSW).

1901: Australia's Constitution is established. The states are empowered to regulate and allocate water.

1902: First Interstate Royal Commission on the River Murray commissioned in May 1902, and expeditiously completed by December 1902.

1914: First intergovernmental agreement (on water) established. The Commonwealth, New South Wales, Victoria and South Australia sign up to the River Murray Waters Agreement, which sets out the shares of water available to each state.

1917: River Murray Commission established.

1918: After World War I, Australian governments create soldier settlement schemes on the Murray to provide jobs and livelihoods for returned servicemen.

1922: First Murray river lock built. Thirteen more will be built by 1939.

1936: Hume Dam completed.

1954: Budj Bim (then called Lake Condah and its swamp system), which survived for thousands of years to feed Indigenous people, is drained. 'They knew what the key to draining the lake was because they had the old surveyor records from the 1890s, where one of our old fellas had told them about the cultural weir. That weir is the one we talk about 4,600 years ago—the key to managing water along the lake and into the systems,' said Damein Bell.

1960s: Chowilla Dam proposed on the Murray River—eventually to be rejected in favour of Dartmouth Dam.

1967: South Australia freezes irrigation entitlements. First study by River Murray Commission into irrigation, drainage and salinity.

1968: Drought causes Hume Dam to drop to 1 per cent of capacity. Menindee Lakes water storage scheme is completed.

1968: Blowering Dam completed as part of the Snowy scheme, a major piece of post-war infrastructure. The scheme affects the Basin in many ways, including by diverting water from the Snowy River to the Murray and the Murrumbidgee.

1979: Dartmouth Dam—becoming Basin's biggest storage—is completed.

1981: Murray mouth closes for first recorded time.

Liberal Prime Minister Malcolm Fraser meets with the Basin premiers to agree on extending the role of the River Murray Commission to cover water quality and other matters. This leads to the first Murray–Darling Basin Agreement in 1982 and the beginning of total catchment management.

1983: The *Murray–Darling Basin Act 1983* is ratified and passed by four parliaments.

1995: (Interim) caps introduced to Murray–Darling Basin.

1997: Millennium Drought starts.

1998: Murray Lower Darling Rivers Indigenous Nations (MLDRIN) established in solidarity to protect Country and Culture. Today

MLDRIN has grown to include 25 Nations. 'Those mobs didn't wait for the invitation,' said Damein Bell.

2000: Murray mouth needs dredging again.

2003: Murray–Darling Basin states agree to find more flows for the Murray.

2004: Council of Australian Governments establishes the National Water Initiative, which elevates the sustainable use of water resources.

2005: The Water Act (Victoria) amendment of 2005 establishes the Environmental Water Reserve, to legally protect water for the environment.

2006: Lowest recorded Basin inflows: 9,300 gigalitres. Around this time, the banks and quasi-banks begin implementing the Holy Grail strategy to corner the water market.

2007: Gavin Jennings, the then Victorian minister for Aboriginal affairs, formally hands ownership of the Budj Bim site to the Gunditjmara people.

The *Water Act 2007* is a high-water mark of bipartisanship in the Basin.

2008: The Australian Human Rights Commission's 2008 *Native Title Report* acknowledges Indigenous people have historically been excluded from water management.

2010: Northern Murray–Darling Basin Aboriginal (NBAN) group formed.

In a moment of peak frustration over the water market and Basin water policy in general, people at Griffith in New South Wales burn hundreds of copies of a Murray–Darling Basin Authority report.

Budj Bim restored and lake starts to fill.

2012: Murray–Darling Basin Plan becomes law, binding all the Basin states and the Commonwealth. The Plan's price tag reaches $13 billion.

2014: UN Intergovernmental Panel on Climate Change (IPCC) releases its fifth assessment report and confirms that warming will continue through the 21st century. Based on projections for warming at 2°C, south-eastern Australia may experience a 40 per cent decline in annual water runoff.

1 July 2014: Water trading restrictions are relaxed.

24 February 2016: Federal government appoints Mick Keogh as a new ACCC commissioner.

August 2017: Satellite imaging and other spatial data proposed to be used to reveal cases in which actual water use does not appear to be consistent with the user's annual 'water return'—akin to tax returns.

21 March 2018: 'Monty Pythonesque' $80 million water buyback in Queensland is 25 per cent more than the asking price. Despite the water being valued at $1,200–$1,500 a megalitre, vendors originally insisted on $2,200 a megalitre; after negotiation the federal government pays $2,750 a megalitre. 'If the water ever flows off the property, it can be captured by adjoining properties, such as Cubbie Station,' Maryanne Slattery, senior researcher at the Australia Institute, says. 'So, it is unlikely that the water will be used for the environment at all.'

30 June 2018: Register of Foreign Ownership of Water Entitlements shows 10.4 per cent (4,035 gigalitres) of all water assets are held by foreign entities. China and the United States are the largest foreign water holders, each with 1.9 per cent, followed by the United Kingdom with 1.1 per cent.

Summer 2018–19: Three separate mass fish deaths occur in the space of two months at Menindee on the Lower Darling. More than a million fish are estimated to have been killed.

29 January 2019: South Australian Royal Commission findings. An important observation from the review is that the use of meters in South Australia was 96 per cent compared to between 25 per cent and 51 per cent in NSW, which was hardly surprising given that water theft allegedly occurred primarily in the latter Basin state.

4 March 2019: Water scientists John Williams and Quentin Grafton estimate that at least 280 billion litres may have been lost from the rivers and is unaccounted for.

22 March 2019: Northern NSW cotton farmer Anthony Barlow is fined $190,000 for breaching *Water Management Act 2000*. Barlow is estimated to have moved the equivalent of almost 153 Olympic swimming pools from the Barwon River to his property near Mungindi on the NSW–Queensland border. In his own defence, Barlow submits that the former NSW Nationals water minister Kevin Humphries had said 'pump unless told otherwise'.

July 2019: Bureau of Meteorology advises that the Murray–Darling is in the 'most severe [drought] condition in 120 years of records'. The Federation Drought (1891–1903) and the World War II drought (1937–1945) were similar to current conditions. Furious irrigators descend on Parliament House wanting to tear up the Murray–Darling Basin Plan. It has been six years since the Plan became law and they still want to burn it.

1 August 2019: Former federal police commissioner Mick Keelty, backed by federal water minister Littleproud as 'tough cop', is installed as interim inspector-general to oversee the Murray–Darling

Basin as part of a government bid to crack down on water theft and other water use breaches.

7 August 2019: Federal government announces that it will direct the ACCC to conduct an inquiry into markets for tradable water rights in the Murray–Darling Basin.

2 September 2019: Minister Littleproud holds a media conference in Echuca following a tour of the Barmah Choke on the Murray River. 'We are now going to be able to provide information in real time, in who owns the water, where the water is, where the allocations are, what's available, who has them,' says Littleproud.

17 October 2019: Professor Sue Jackson of Griffith University observes it wasn't until the mass fish kills in Menindee Lakes that politicians and officials sat up and took notice. Despite thousands of years of knowledge, Indigenous Australians have been relegated to stakeholders whose approval needed to be 'ticked off' rather than custodians and water managers with unique and vital skills. Reporter Willow Aliento writes: 'Jackson told *The Fifth Estate* the crux of the matter is 20 years or more of unsustainable water extraction from the top of the catchment by the big irrigators…Aboriginal people in Northern NSW have been talking to scientists and historians for years about the state of the waterways, Jackson says. They reported rivers running backwards because of the volumes being extracted for irrigation and other purposes.'

15 December 2019: NSW town of Evans Plains robbed of 300,000 litres of water during drought and the blackest summer ever. While there might be something morbidly entertaining about the water theft—particularly since it has a very *Mad Max*–style dystopia feel in the country where that series was filmed—it's far from a laughing matter. In fact, it's not even the first such heist to hit the region.

A report from the *Sydney Morning Herald* on 12 December 2019 documented a similar scheme that led to 25,000 litres of drinking water being siphoned from the small town of Murwillumbah, which is currently under a water restriction as government officials try to stave off the threat of the town going completely dry.

January 2020: Reports show that Australian rainfall in 2019 was 40 per cent below average, and the Murray–Darling Basin river system had its driest year on record.

30 July 2020: The ACCC releases its Murray–Darling Basin water markets inquiry interim report, which draws upon analysis of comprehensive water market data from 2012 onwards, and reflects the views of a broad range of people with interests in the use and trading of water in the Basin.

4 September 2020: The Commonwealth government announces it will stop water buybacks in the wake of the Sefton report on social and economic conditions in the Basin.

30 October 2020: Mick Keelty resigns after a year in the job as the Basin's 'top cop'.

23 November 2020: A University of NSW study finds that the area occupied by platypuses in the Basin has declined by 30.6 per cent since 1990. Their habitat has been badly affected by over-extraction of water from rivers, land clearing, pollution, urban sprawl, regulation of the rivers and drought. Richard Kingsford, one of the study's lead authors, tells the *Guardian*: 'There is a real concern that platypus populations will disappear from some of our rivers without returning if rivers keep degrading with droughts and dams.'

27 November 2020: NSW joins with Victoria at a meeting of state water ministers to call for a review of the Murray–Darling Basin

Plan in 2022. South Australia opposes this, and the Commonwealth blocks the move. Lisa Neville, who has been Victoria's water minister for six years, says, 'Today was the first time I've seen the Commonwealth step in and say, 'No, we're not going to accept that compromise."'

4 December 2020: Treasurer Josh Frydenberg asks the Productivity Commission to inquire into the effectiveness of the Register of Foreign Ownership of Water Entitlements.

18 December 2020: The Murray–Darling Basin Authority welcomes the appointment of Nari Nari man Rene Woods to the permanent Indigenous Authority Member board position.

26 March 2021: ACCC final report released. ACCC's Mick Keogh says there are a heap of problems with Australia's $2 billion water market, including that market manipulation is not illegal.

The future: Strictly speaking, in an elemental sense, there is no shortage of water; just water in the wrong place, and of the wrong type. With infinite, cheap, non-polluting energy, seawater can be turned into freshwater and moved to wherever it is needed. But we don't have infinite, cheap, non-polluting energy, so we have to make the best of where we are at, and where we are likely to be for some time, hydrologically and technologically.

'With a growing population,' farmer Shelley Scoullar says, 'we want to protect our environment. How can we do that sustainably? How can we work together? Along the way, I've come across reconciliation ecology. I really think Australia is a great country where we can lead the world in having ecological outcomes in human-dominated spaces. If we think outside the square, I know it can happen.'

APPENDIX: WATER WORDS

Words are a large part of the problem in water. They help explain why governments over recent decades have taken the wrong path; and they are one reason why traders have been able to get away with murder. Traders use words strategically to obscure their tactics, justify their activities and influence the market.

Word choices are also a visible aspect of water traders' learning and innovation. The banks and traders are continually coming up with new 'water investment products' and new stories to wrap around their water market activities. The water-related funds and exchanges are running out of ways to use 'water': Waterfind, The Water Fund, Waterflow…

(When Enron was being established, its founders preferred the name 'Enteron', until they discovered it meant 'intestines'. Our favourite alternative name for the Holy Grail strategy is 'Wateron'.

In 2020, in a move that was aptly described as involving 'Monty Python Latin', resources company Adani renamed its Australian operations 'Bravus', which means 'crooked' rather than 'brave'.)

Often in unhelpful ways, the water policy lexicon blends terms from different disciplines: economics, finance, politics, public policy, hydrology, engineering, environmental science, community development. Problems of language are problems of conceptual confusion, hidden assumptions and concealed ideology.

As Riverina farmer Lachlan Marshall noted, the lexicon is full of jargon and euphemisms, and the vocabulary is highly fluid.

'Things are always being given new names,' he said, 'like for different water products or actions. One that really got under my skin is "over bank transfer", which is the code-word for a flood. "We couldn't fit it down the river so we jammed it over the bank!"'

When we examined the past 20 years of water policy reports, we found many instances of terminological confusion and fuzzy analysis. Water, we think, is too important to be left in such a verbal mess that confuses everyone. Here, in the interests of helping to reduce the confusion, is our water markets lexicon.

Efficient, efficiency: These words are highly prominent in the water policy debate, and especially in contributions from the ACCC, the Productivity Commission and other public policy defenders of neoclassical economics. According to the PC, for example, 'The NWI aims to facilitate efficient water markets.' But here, too, there is linguistic and disciplinary confusion.

In economics, 'efficiency' has specific meanings relating to resource allocation, innovation and the avoidance of waste. Economically speaking, a business achieves 'technical efficiency' if it maximises output for a given level of inputs. 'Allocative efficiency' is where products and resources are allocated to the people who value them the most. 'Dynamic efficiency' means that, over time—with innovation and the reallocation of resources—production and resource allocation continue to be (technically and allocatively) efficient.

The discipline of finance introduces other meanings of 'efficient', which complicate conversations about markets and efficiency. In finance, the 'efficient markets hypothesis' is the idea that prices in a well-functioning market will reflect all available information, and therefore the current price is the best indicator of the future price. In the financial sense, therefore, Australian water trading is most certainly not an 'efficient market'—there is systemic incorrect pricing of water because information is not shared transparently. ('Efficient water market' also has another, everyday meaning. It refers in a general sense to whether the market functions well: whether transaction costs are low, the market processes are not onerous, and so on.)

Allocative efficiency is the variety of economic efficiency that is most relevant to water trading. This is about whether the water market tends to set prices and enable transactions that support the movement of water rights to where they are most valued. The ideas of allocative efficiency (economic) and efficient markets (financial), though related, are sufficiently different that, next time someone comments on the efficiency of water markets, you need to question them further as to what they are talking about.

By 'efficient water markets', the PC mainly means 'markets that improve allocative efficiency', not the financial markets concept of 'water prices use all available information and are therefore the best predictor of future prices', though aspects of the latter (such as generally well-functioning markets) are embedded in the PC's words. Even the ACCC has misused the phrase 'efficient market' from time to time.

Has the design of our water market increased allocative efficiency? In some respects, the inclusion of outside participants may have tilted the market *away* from allocative efficiency in a number of ways: by pushing prices above an allocatively efficient level; by discouraging on-exchange water trading, and therefore muting price signals; by reducing confidence in the market; and, as we found, by allowing those participants to game the market and to send false signals that allocate water in an unsustainable and perverse way.

Is the water market 'efficient' in an everyday sense? According to our journey (and as confirmed by the ACCC), the answer is no. Our trading environments are too incomplete and dysfunctional. Many trades are not correctly reported, markets are disconnected, and participants rightly doubt the integrity of some markets and some market conduct. Information is often stale, of poor quality and unevenly dispersed.

Financialisation: 'Financialisation' involves turning non-financial products and services into tradable, financial assets. Energy spot markets and energy futures markets were prominent early examples.

Hedge funds: These are large funds that deploy agile, smart money, after searching the world for high returns, usually from complex arbitrage opportunities within and across financial markets. (The line between hedge funds and private equity funds is distinctly blurred, as is the line between hedge funds and investment banks.) Note that 'hedge funds' have very little to do with 'hedging', just as 'short selling' is not about shortages or scarcity.

Investment, investing: As we have shown, the Basin water market actually comprises multiple markets. As with 'efficient markets', when someone says they are 'investing in water', your first response should be, 'What precisely do you mean?' There are multiple trading platforms, and there are regional markets, different types of water, different types of water rights, and different types of irrigation investment. Some investors are buying shares in irrigated land. Some are buying permanent water rights, as a long-term nest egg. And some are actively trading water, day-to-day.

The terms 'investment' and 'investing' are therefore especially problematic in water. They are umbrella terms for lots of different behaviours, some of them benign, some predatory. Assembling and applying superannuation funds in order to build an olive grove, in the hope of a return in ten years' time, is a routine and generally benign form of investing. But the early movers in the water market found they could invest in a much more lucrative way if they played the market.

In economics, 'investment' means the produced means of production: making and improving capital equipment and other physical things, such as farms, that can produce goods and services. In other contexts, the word 'investment' is often used to mean 'saving', such as when someone 'invests' in a managed fund.

When big traders buy water rights, they are often doing neither of these things: they are certainly not investing in the economic sense; and mostly they are not saving. External traders operate more like hedge funds, buying and selling water rights in ways that make a trading profit. If they wanted somewhere to save, there are many alternative options.

Investment banks: These organisations have little interest in investment, and they are not 'banks' in the conventional understanding of that word. They are quite like hedge funds, and they are exactly like money factories.

Liquidity: Apart from being metaphorically relevant to water trading, 'liquidity' has a specific meaning in finance. It means there is depth and continuity in the market: there are multiple buyers and sellers, who make a range of bids and offers, and they do so even when markets are under pressure or in crisis. As a result, the 'bid–ask spread' is reliably narrow.

Do outsiders in the water market help increase liquidity in this sense? Only sometimes. They may in fact reduce liquidity by holding on to rights, or offering them only at prices that are 'out of the money', in order to affect the market and ultimately to sell the rights at a higher price.

In some of the official policy papers and regulatory reports, the word 'liquidity' is redefined and reimagined downward to mean simply 'more participants' in the market. In a limited and direct sense, allowing external participants to trade in water can increase the number of market participants. However, the activities of external players—driving up prices, disentangling water values from the 'value in use'—can impair the functioning of markets, and it can otherwise crowd out other participants. So the impact on 'liquidity' is far from straightforward.

Private equity: Another noticeable feature of the new financial world. Cashed-up corporate raiders bought household-name businesses and sold them quickly, earning a 30 per cent return so new owners could run the restructured business for a 10 per cent return. The term 'private equity' is misleading; it is more about debt than equity. Private equity transactions are what we used to call, in the 1980s and 1990s, 'leveraged buy-outs' or LBOs—where 'leverage' means 'debt'.

Risk management: A striking feature of today's water world is the growth of markets in water-related financial assets such as water futures, water bonds and water options. In that growth, we see a significant opportunity for the Murray–Darling Basin. Part of the problem can become part of the solution.

The new asset types can be used for legitimate risk management, and they offer a means to purge the gambling element from the water market itself. That element could be quarantined in side markets, where the traders could gamble all they liked, in an environment safely decoupled from the distribution and use of water. Synthetic securities and derivatives could expand the availability of water-linked tradable assets without perverse effects on the allocation of a scarce resource. And they would help introduce proper risk management into Australian water.

Short selling: Conventional short selling is a kind of bet: a gamble that a stock or some other type of asset will fall in price. The aim is to 'borrow' an asset that you don't own, sell it at a high price, then buy it at a lower price to return the asset to its owner, while keeping the profits. Not all securities are approved for short selling. In Australia, the shares you can 'short' are typically the frequently traded ones and those of larger companies.

Investment fund Blue Sky Alternative Investments Ltd, which owned water assets and much else besides, was famously the subject of an attack by 'activist short sellers' (who undertake and publish research that drives down the value of the companies they are attempting to short). In March 2018, short seller Glaucus claimed Blue Sky had overstated its fee-earning assets under management, and that Blue Sky shares were more than 75 per cent over-valued.

Speculation, speculating: In the popular understanding, 'investing' is good, and 'speculation' is bad. People associate speculation with predatory, anti-social behaviour, and with betting and gambling.

Generally, however, to 'speculate' simply means to 'make judgments and decisions under uncertainty'—something all water market participants do, not just traders. And some water market speculation is not a gamble at all: it entails little or no risk.

As with 'efficiency' and 'investing', 'speculation' encompasses lots of different short- and long-term water market phenomena. For example, it encompasses the actions of traders hoping to 'buy low and sell high' to turn a profit. It also includes taking a longer term position in the market, in the speculative expectation that water rights will become much more valuable, and therefore deliver a big, long-term gain.

Some speculation can be helpful in markets; it can help people manage risk and seek useful market exposures. An exposure to the value of water, for example, can be helpful for setting climate change–related incentives and for hedging risk. These 'good' forms of speculation can be accommodated without disrupting the allocation of water rights.

The problem in the water market is not so much about speculation and speculators; it is more about gouging, skimming, rent-seeking, gaming, monopolising and cornering.

ACKNOWLEDGMENTS

Most of the events described in this book took place on the lands of 40 First Nations, whose rich culture and traditions stretch back more than 70,000 years. We acknowledge the Traditional Custodians and their continuing and unique connection to the land, waters and community. We pay our respects to all members of Aboriginal communities and cultures; to their Elders past, present and emerging; and to their fellow Indigenous people near and far.

The modern history of Indigenous Australians is marred by crimes and injustices, including theft and dispossession of Indigenous water rights. Genuine reconciliation requires a national process of truth-telling, plus a constitutionally enshrined voice in the parliament—as set out in the Uluru Statement from the Heart—and justice and voice regarding water rights and water management.

The research for this book was very much a collaborative effort.

We listened with gratitude to Shelley Scoullar, Lachlan Marshall, Laurie Beer, Peter and Renee Burke, Ben Dal Broi, Chris Brooks, Damein Bell, David and Jeanette Crew, Keith Hamilton, Kate Auty, Ian Shepherd, Paul Pierotti, Robyn Wheeler, Tony Windsor and Malcolm Turnbull. Shelley Scoullar in particular was incredibly generous in sharing her personal story and her Basin connections.

Some interviewees spoke to us on condition of anonymity. Accordingly, those interviewees appear in the text as 'Mr X' and 'Mr Y' (for anonymous water exchange managers) and 'Shane Sutherland', 'Desiree Button', 'Clinton Davis' and 'Gary Weber' (for anonymous brokers and traders). If you are an exchange manager, a water trader or a water broker and you recognise yourself in this book, that is only natural, and very likely coincidental. We also acknowledge the generous and valuable input of several anonymous mentors and reviewers.

We are grateful to the librarians and other staff of the state libraries of Victoria, New South Wales, South Australia and Queensland; the National Library of Australia; Riverina Regional Library; Federation University, Ballarat; La Trobe University; Charles Sturt University; Griffith University; the University of Melbourne; and the Australian National University.

Apart from being a collaborative project, this book was also a family project. Our families, including the Hamiltons' Sarah, Astrid and Atticus and the Kellses' Fiona, Thea and Charlotte, supported us tirelessly with reading drafts, transcribing interviews, accommodating countless Zoom calls, bringing cups of tea and coffee (and water), and supplying more than the occasional piece of chocolate cake. Fiona Kells helped find key documents, assisted with referencing, reviewed multiple drafts of the manuscript, and otherwise assisted in dozens of practical and essential ways.

We are grateful to Ben Ryder Howe and the *New York Times* for featuring our research and highlighting the important connections between the Australian and American water markets. Australia's epic fail is a cautionary tale for US states such as Colorado and California. We also thank Colorado attorney-general Phil Weiser for his remarks at the Colorado Water Congress 2021 Annual Convention, where he spoke on 'The Right and Wrong Ways to Manage Water: Lessons from Spectrum and Australia'.

The amazing editorial and media teams at the University of Melbourne's *Pursuit*, the *Mandarin*, La Trobe University and more recently *Ticker NEWS* have been invaluable in helping us share the results from our research.

We acknowledge the prior work of other scholars, authors and journalists, including Michael Lewis, Michael Cathcart, Ernestine Hill, Chris Hammer, Gabrielle Chan, Richard Beasley, Kate Jennings, Tim Flannery, Sue Jackson, Quentin Grafton, Maryanne Slattery, Bill Arthur, Rene Woods, Phil Duncan and Margaret Simons.

It was a pleasure to work with the marvellous team at Text Publishing, including Michael Heyward, Ian See, Sophie Mannix, Anne Beilby, Lara Shprem, Stefanie Italia, Kate Lloyd, Jane Watkins, Emily Booth and our magnificent cover designer, Jessica Horrocks. We are also grateful for the work of our brilliant cartographer, Simon Barnard, and our astute proofreader, Nikki Lusk.

NOTES

All quotes from the following people were taken from interviews we conducted in the stated months: Laurie Beer, September 2020; Damein Bell, October 2020; Chris Brooks, October 2020; Peter Burke, October 2020; Desiree Button*, October 2020; David Crew, December 2020; Ben Dal Broi, September 2020; Clinton Davis*, October 2020; Lachlan Marshall, September 2020; Paul Pierotti, November 2020; Shelley Scoullar, December 2020; Shane Sutherland*, October 2020; Gary Weber*, October 2020; Robyn Wheeler, November 2020. *(* Pseudonyms were used for these water brokers and traders.)*

INTRODUCTION

'Water provides many...derails bipartisan solutions': Scott Hamilton and Stuart Kells, 'Murray–Darling Basin bipartisanship: not new, not strong,' *Mandarin*, 20 February 2019.

around $1.8 billion per year...value of Basin farms' assets: Australian Competition and Consumer Commission (ACCC), *Murray–Darling Basin Water Markets Inquiry: final report*, Australian Government, 2021, p. 80.

'It's byzantine...Blah, blah, blah.': University of California Television (UCTV), 'Water and Power: discussion of documentary', *UCTV*, YouTube, 15 September 2017.

CHAPTER 1

'Ants building their nests...a wet season', *'We have found...storm is coming'*, *'My dad always...rain was coming!'*, *'Ducks nesting in...Big rain coming'*, *'My mum is convinced...rain two days later'*, *'We've always thought...until it stops again'*: Interview with Shelley Scoullar, December 2020.

'In a good year...next week's weather': Richard Shears, 'The Greenhouse Syndrome', *Australian Playboy*, February 1980, p. 97.

CHAPTER 2

'involves an ancient...and water management': Monica Morgan, Lisa Strelein and Jessica Weir, *Indigenous Rights to Water in the Murray Darling Basin: in support of the Indigenous final report to the Living Murray Initiative*, research discussion paper no. 14, Native Title Research Unit, Australian Institute of Aboriginal and Torres Strait Islander Studies, 2004, p. 17.

'We, the Indigenous...given us everything': Murray Lower Darling Rivers Indigenous Nations (MLDRIN), *Echuca Declaration 2007*, p. 1.

'masters of Nature': Rhys Jones, 'Fire-Stick Farming', *Fire Ecology*, vol. 8, 2012, p. 3. Reprinted from *Australian Natural History*, vol. 16, no. 7, September 1969.

'the very presence of humanity itself': Jim Bowler, quoted in Mungo National Park, *Mungo Lady and Mungo Man*, Mungo National Park website, n.d.

'This research extends...of Indigenous Australians': Jim Bowler, quoted in University of Adelaide, *New Age for Mungo Man*, University of Adelaide website, 9 February 2003.

'We were determined...rights under white law': Interview with Keith Hamilton, September 2020.

'[Gunditjmara Elders] Iris and Charlie...the brolgas were back!': Interview with Kate Auty, January 2021.

'When this went...be jumping for joy': Interview with Kate Auty, July 2019.

According to a Ngemba Wayilwan Dreaming story...thrown over the river: Department of Agriculture, Water and the Environment (DAWE), *National Heritage Places—Brewarrina Aboriginal Fish Traps (Baiame's Ngunnhu)*, DAWE website, Australian Government, n.d.

CHAPTER 3

'the satisfaction of...in the world': Arthur Phillip, dispatch written to Lord Sydney, 15 May 1788, in *Historical Records of New South Wales*, vol. 1, part 2, Charles Potter, Sydney, 1892, p. 122.

'Charles Sturt observed...with birds and fish': Michael Cathcart, *The Water Dreamers*, Text Publishing, Melbourne, 2009, p. 17.

'in case you should...enable you so to do': Matthew Flinders, *A Voyage to Terra Australis*, vol. 1, G. and W. Nicol, London, 1814, p. 8.

'It is impossible to...the heart of the interior': Joseph Banks, letter to John King, 15 May 1798, in *Historical Records of New South Wales*, vol. 3, Charles Potter, Sydney, 1895, p. 383.

'confident we were...a shoal one': Letter to Governor Macquarie, 1 November 1818, in John Oxley, *Journals of Two Expeditions into the Interior of New South Wales*, John Murray, London, 1820, p. 383.

'attractions for the explorer...so bountifully waters': House of Lords, *Sessional Papers*, vol. 5, 1841, p. 13.

'elevated above all...of mountain scenery': Paul Strzelecki, *Physical Description of New South Wales and Van Diemen's Land*, Longman, Brown, Green & Longmans, London, 1845, p. 62.

'looking from the...to the beautiful': ibid.

CHAPTER 4

'the tragedy of the commons': William Forster Lloyd, *Two Lectures on the Checks to Population*, Oxford, 1833.

'Logic clearly dictates...needs of the few': Dr Spock, in Nicholas Meyer [director], *Star Trek II: The Wrath of Khan*, Paramount Pictures, 1982.

'the men of the trowel rather than the sword': Judith Brett, *The Enigmatic Mr Deakin*, Text Publishing, Melbourne, 2017, p. 176.

'the government needed...too deeply involved': ibid., p. 114.

'A recognition of...be sold separately': Alfred Deakin and Victoria Royal Commission on Water Supply, *First Progress Report: Irrigation in Western America, so far as it has relation to the circumstances of Victoria*, John Ferres, Government Printer, Melbourne, 1885, p. 46.

'in his careful...with land titles': ibid.

'In Colorado...was not entertained': ibid.

'Under Deakin's scheme...small annual fee': Cathcart, *The Water Dreamers*, p. 254.

'a fourth-rate pressman...a plausible tongue': Ernestine Hill, *Water Into Gold*, Robertson & Mullens, Melbourne, 1954, p. 64.

'Deakin met with...negotiated an agreement': Brett, *The Enigmatic Mr Deakin*, p. 121.

'repelled by the...frontier individualism': Cathcart, *The Water Dreamers*, p. 203.

'Yankee land grab': Tommy Bent, quoted in Brett, *The Enigmatic Mr Deakin*, p. 121.

'secure the application...in practical irrigation': Hill, *Water Into Gold*, p. 68.

'George Chaffey was...rolling past every year': ibid., p. 67.

'The main issue...prospects for Mildura': Brett, *The Enigmatic Mr Deakin*, p. 174.

'He believed he...boots for a pillow': ibid., p. 176.

'deeply disillusioned by...engulfed the colony': Judith Brett, 'Lessons from Deakin: Alfred Deakin and the art of minority government', *The Monthly*, 7 October 2016.

'gang of swindlers': Howard Frederick, 'Australia's spiritualist water entrepreneur...and prime minister', *entreVersity*, 26 July 2019.

'He argued in...fruit to market': Brett, *The Enigmatic Mr Deakin*, p. 214.

CHAPTER 5

'The Commonwealth shall not...conservation or irrigation': Australian Constitution, section 100.

'In the absence ...interstate rivalrous self-interests': Murray–Darling Basin Royal Commission, *Report*, South Australian Government, 2019, p. 38.

CHAPTER 6

'the corporatisation of agricultural land and water rights': National Water Commission (NWC), *Water Markets in Australia: a short history*, Australian Government, 2011, p. 28.

'a farmer would pay...the next two weeks': Department of Natural Resources and Environment, *The Value of Water: a guide to water trading in Victoria*, Victorian Government, 2001, p. 7.

'If you said...slab of beer': George Warne, in Ticky Fullerton [host], 'Sold Down the River', *Four Corners*, ABC TV, 14 July 2003.

'Trading can reveal...most highly valued': Productivity Commission, *Rural Water Use and the Environment: the role of market mechanisms*, research report, Australian Government, 2006, pp. 67–68.

'The government solution...bad as the problem': Milton Friedman, *An Economist's Protest*, Thomas Horton & Co., Glen Ridge, 1972, p. 6.

'The primary mechanism...adjustment in rural industry': Australian Water Resources Council, *Proceedings of the Joint AWRC–AAES Seminar on Transferable Water Rights*, Australian Government Publishing Service, Canberra, 1986, p. v.

'The "unbundling" of…their individual enterprises': Department of Sustainability and Environment, *Securing Our Water Future Together*, Victorian Government, 2004, p. 69.

'At the moment…owners logged there': Russell Cooper, quoted in 'Irrigators to be told of water entitlements "unbundling",' *ABC News*, 7 March 2007.

'rights to water…unbundled from land': Aither Pty Ltd, *Review of Water Trading: the impact of IPART's regulatory framework on water trading markets: a final report prepared for IPART*, 2018, p. 6.

CHAPTER 7

'The ACCC notes…for the environment': ACCC, *Water Trading Rules: final advice*, Australian Government, 2010, p. xii.

'there should be…use of water', *'there should be…Murray–Darling Basin'*, *'trading rules…water carried over'*, and *'trade between regulated…or hydrological issues'*: Australian Bankers' Association, submission in response to ACCC, *Water Trading Rules: draft advice*, 2010, pp. 1–3.

'There are now…in the meantime': Danny O'Brien, quoted in NWC, *Water Markets in Australia*, p. 90. Also accessible at *The 4 per cent rule—help or hindrance?*, National Irrigators' Council website, 19 January 2010.

'There is concern…actually using it': Victorian DSE, *Securing Our Water Future Together*, p. 69.

'All the permanent…business to another': ibid.

'The non-water user…non-water user limit': Victorian Legislative Assembly, *Debates*, 29 July 2009, vol. 9, pp. 2391–92.

'where investors have…remains in the market': ibid., p. 2392.

'ABA agrees that…to be landowners': Australian Bankers' Association, quoted in ACCC, *Water Trading Rules: final advice*, pp. 35, 45.

'The Basin Plan water…Foreign Acquisitions and Takeovers Act 1975)': ACCC, *Water Trading Rules: final advice*, p. 40.

'Actually there really…would look like': Interview with a water exchange

manager, November 2020.

'Australia's experience of...other natural resources': NWC, *Water Markets in Australia*, p. xiii.

CHAPTER 8

'and therefore only...given to irrigation': Henning Bjornlund and Brian O'Callaghan, 'Property implications of the separation of land and water rights', *Pacific Rim Property Research Journal*, vol. 10, no. 1, January 2003, p. 7.

'In order to protect...prevent environmental degradation': United Nations General Assembly, *Report of the United Nations Conference on Environment and Development, Annex I, Rio Declaration on Environment and Development*, A/CONF.151/26 (vol. I), 12 August 1992, p. 3.

'holders of such...aggregate water use': NWC, *Water Markets in Australia*, p. 43.

'Australia's largest river...region from annihilation': Cathcart, *The Water Dreamers*, p. 253.

'Overallocation of water...of water extraction': National Water Commission, *National Water Initiative: first biennial assessment of progress in implementation*, Australian Government, 2007, p. 3.

The inaugural chair...managing Basin water resources: Peter Ker, 'Murray–Darling chair resigns, leaving reforms in turmoil', *Sydney Morning Herald*, 7 December 2010.

'We have had...Adelaide's water requirements': Australian Senate, *Debates*, 17 September 2008, vol. 9, pp. 4986–87.

'almost became paranoid...frailty in it': Tony Windsor, quoted in Scott Hamilton and Stuart Kells, 'The bipartisan kingmaker: an interview with Tony Windsor', *The Mandarin*, 26 August 2020.

'very poor', *'inland water ecological processes and key species populations'*: Robert Argent, *Australia State of the Environment 2016: inland water*, Australian Government, 2016, p. 56.

'This has nothing...a man-made disaster', *'You have to...and cotton growers'*: Rob McBride and Dick Arnold, quoted in Peter Bannan, 'Fish Kill', *Guardian* (Swan Hill), 21 January 2019.

'The conditions leading...lack of local reserves': Australian Academy of Science (AAS), *Investigation of the Causes of Mass Fish Kills in the Menindee Region NSW over the Summer of 2018–2019*, AAS, 2019, p. 2.

'was bipartisan until...under a bus': Tony Burke, quoted in Laura Tingle, 'The fury of the Murray–Darling royal commission findings has fallen on deaf ears', *ABC News*, 2 February 2019.

'It took us...to be some tweaks': Michael McCormack, quoted in Caroline Winter, 'Deputy Prime Minister signals room to tweak the Murray Darling Basin Plan', *The World Today*, ABC Radio, 17 January 2019.

'It seems to be...success of the Plan': Steve Whan, quoted in Kath Sullivan, 'Labor plans to scrap cap on water buybacks across Murray–Darling Basin', *ABC News*, 12 February 2019.

'a very ill patient...out of the ICU': Julie Power, 'Macquarie marshes: a great sight of nature only being kept alive by life support', *Sydney Morning Herald*, 19 November 2016.

'It's a place...It's dead water': Lilliana Bennett, quoted in Aneeta Bhole, 'Indigenous community say they've lost their culture to water mismanagement', *SBS News*, 18 October 2019.

CHAPTER 9

'The reformers talked...along the high street': Cathcart, *The Water Dreamers*, p. 257.

'A lot of...into your crop': Ange Lavoipierre and Stephen Stockwell [hosts], 'Australia's water barons', *The Signal*, podcast, ABC Radio, 8 October 2019.

'In our district...get them wet': Frontier Economics et al., *The Economic and Social Impacts of Water Trading: case studies in the Victorian Murray Valley*, report for the Rural Industries Research and Development

Corporation, National Water Commission and Murray–Darling Basin Commission, 2007, p. 136.

'*Water trade has...easily at the moment*': ibid., pp. 74–75.

'*the unbundling of...may fall significantly*': Michael Woolston, quoted in NWC, *Water Markets in Australia*, p. 67.

'*There is a chap...had a go*': Frontier Economics et al., *The Economic and Social Impacts of Water Trading*, p. 183.

'*Years ago, I probably...hard to keep positive*': ibid., p. 132.

'*a general sense of grieving*': ibid., p. 160.

'*The whole thing...the processing works*': ibid., pp. 160–61.

'*I've got a mate...him going down*': ibid., p. 75.

'*At the moment ... we are used to*': ibid., p. 77.

'*The biggest problem...a bloody disaster*': ibid., p. 161.

'*Because of permanent...It upsets people*': ibid., p. 194.

'*Yes we have...sold that much*': ibid., pp. 73–74.

CHAPTER 10

'*We own the...and the structures*': Frontier Economics et al., *The Economic and Social Impacts of Water Trading*, p. 138.

'*the Tinder of the water industry*': Mark Veitch, in NSW Legislative Council, *Report on Proceedings Before Portfolio Committee No. 5—Industry and Transport: Water Augmentation*, 5 June 2017, p. 5.

'*Water literacy has...farms falling behind*': Mick Keelty, quoted in Kath Sullivan and Clint Jasper, 'Mick Keelty's Murray–Darling Basin review finds declining inflows, toxic debate and need for education', *ABC News*, 17 April 2020.

'*Just because you have...they can't fill it*': Lavoipierre and Stockwell [hosts], 'Australia's water barons', *The Signal*, 8 October 2019.

'*We're seeing now...$2,000 to $2,200*': Nikolai Beilharz [host], *Victorian Country Hour*, ABC Radio, 26 November 2019.

'So if we...different prices there': ibid.

'You have to be...what I've been doing': Frontier Economics et al., *The Economic and Social Impacts of Water Trading*, p. 75.

CHAPTER 11

'Water is the oil of the 21st century': Cited by Lynne Kirby in University of California Television (UCTV), 'Water and Power: discussion of documentary', *UCTV*, YouTube, 15 September 2017.

'As the global...business opportunities': Cathcart, *The Water Dreamers*, p. 256.

'elitist multibillionaires'; *'all over the world at [an] unprecedented pace'*: Jo-Shing Yang, 'The new "water barons": Wall Street mega-banks are buying up the world's water', *Market Oracle and Global Research*, 21 December 2012.

'There is no depreciation...that are that good': Constantin Seidl, Sarah Ann Wheeler and Alec Zuo, 'Treating water markets like stock markets: key water market reform lessons in the Murray–Darling Basin', *Journal of Hydrology*, vol. 581, February 2020.

'They raced to it like 17-year-olds at the Cootamundra B&S Ball': Interview with a water broker, August 2020.

'Yes they are...They were invited': Interview with Mr X, March 2020.

'cheerleaders': ibid.

'Burn, baby, burn...two a day': Quoted in Joel Roberts, 'Enron Traders Caught on Tape', *CBS News*, 1 June 2004.

'The banks have...understand this market': Interview with Mr Y, March 2020.

CHAPTER 12

grew by an amazing 2 million per cent: 'Capital markets explained', *Chambers Student*, February 2011.

Three of Australia's...between 2010 and 2012: Stephen Letts, 'ANZ, NAB agree to $100 million settlement of swap rate rigging case', *ABC News*, 10 November 2017.

In 2019, Australian legal action...businesses and investors: David Chau, 'Citibank, UBS, JP Morgan, Barclays and RBS sued for rigging currency exchange rates', *ABC News*, 27 May 2019.

In another recent case...fourth criminal case since 2012: Tom Schoenberg, Matt Robinson and Bloomberg, 'JP Morgan will pay record settlement to resolve "spoofing" case against 15 traders', *Fortune*, 30 September 2020.

As Australia's financial regulator ASIC noted...new derivative products: Australian Securities and Investments Commission, *Mandatory Central Clearing of OTC Interest Rate Derivative Transactions*, consultation paper 231, Australian Government, May 2015.

CHAPTER 13

'*It's an incredibly...in the market*': Mick Keogh, quoted in Mike Foley, '"Serious distrust" in murky Murray Darling water market sinks farmers', *Sydney Morning Herald*, 30 July 2020.

'*A few [of the traders] are older...most are young turks*': Interview with a water market insider, August 2020.

'*To picture the...of play money*': Interview with a water market insider, December 2020.

'*transact millions of...tax file number*': Keogh, quoted in Foley, '"Serious distrust" in murky Murray Darling water market sinks farmers', *Sydney Morning Herald*, 30 July 2020.

'*The banks just...At least at first*': Interview with a water trader, October 2020.

NSW Water's 15 October 2020 announcement...on the MIL exchange: Email communication from Laurie Beer, 16 October 2020.

'*We are about to...water market price*': Beilharz [host], *Victorian Country Hour*, 26 November 2019.

CHAPTER 14

'*In October 1997...Soros's wildest dreams*': Niall Ferguson, *The Ascent of Money*, Allen Lane, London, 2008, p. 325.

'*We saw recently...got their water out*': Beilharz [host], *Victorian Country Hour*, 26 November 2019.

'*I suspect they don't publish it because it would be terrible!*': Interview with a water broker, August 2020.

'*In almost all...a bid at $180/ML*': Argyle Capital Partners, submission in response to ACCC, *Murray–Darling Basin Water Markets Inquiry: interim report*, 2020.

CHAPTER 15

'*the "fastest finger" wins when trading windows open*': goFARM Pty Ltd, submission in response to ACCC, *Murray–Darling Basin Water Markets Inquiry: interim report*, 2020, p. 3.

'*Any time the Choke...it's driven by bots*': Interview with a water trader, November 2020.

'*It's amazing we're...about the water*': Interview with Mr X, March 2020.

Financial author Edgar Peters...over four-year cycles: Edgar Peters, *Chaos and Order in the Capital Markets: a new view of cycles, prices, and market volatility*, Wiley, California, 1991.

Benoit Mandelbrot found...market for cotton: Benoit Mandelbrot, 'The variation of certain speculative prices', *Journal of Business*, vol. 36, no. 4, October 1963, pp. 394–419.

'*No question that...access to market metadata*': Email communication from a Basin businessman, February 2020.

'*I'm in a difficult position...It's opportunistic*': Frontier Economics et al., *The Economic and Social Impacts of Water Trading*, p. 194.

'*It's their own...and seller beware*': Interview with a water trader, November 2020.

'*I'm not entirely...get with the times*': ibid.

CHAPTER 16

'*Inter-valley trading is...rorting takes place*': Interview with a water market insider, November 2020.

'*by misrepresenting the...actually the case*': ACCC, *Murray–Darling Basin Water Markets Inquiry: interim report*, Australian Government, 2020, p. 201.

'*Information in the...if any consequences*': Interview with a water exchange manager, March 2020.

'*You may be interested...providing misleading information?*': Email communication from Laurie Beer, 16 October 2020.

CHAPTER 17

'*You're not going to believe my answer*': Interview with a lawyer, December 2020.

'*the Australian Securities...as a financial market*': Explanatory Statement, Corporations Laws Amendment (2014 Measures No. 1) Regulation 2014: Water Trading Exemptions, p. 1.

'*Back in 2004...to sell them*': Tom Rooney, quoted in Rod Myer, 'Water could be gold for Australian superannuation fund members', *New Daily*, 23 March 2019.

'*to be recognised...borrow against them*': ibid.

'*reduce borrowing costs...directly over water*': Victorian DSE, *Securing Our Water Future Together*, p. 69.

'*It almost feels like "anything goes" in this market*': Interview with a Basin businessman, November 2020.

CHAPTER 18

'*harvested by financial...ears of corn*': Alan Kohler, *It's Your Money: how banking went rogue, where it is now and how to protect and grow your money*, Nero, Melbourne, 2019, p. v.

'*most financial advisers...culture has endured*': Royal Commission into

Misconduct in the Banking, Superannuation and Financial Services Industry, *Interim Report*, Commonwealth of Australia, 2018, vol. 1, p. 76.

'*the planner always...sell you something*': Kohler, *It's Your Money*, p. v.

'*Experience shows that...always trump duty*': Royal Commission into Misconduct in the Banking, Superannuation and Financial Services Industry, *Final Report*, Commonwealth of Australia, 2019, vol. 1, p. 3.

'*Right from the...of insider trading*': Interview with a water official, December 2020.

'*The regulators and the rules are useless*': Interview with a lawyer, December 2020.

'*It's a bloody mess*': Interview with a lawyer, December 2020.

CHAPTER 19

'*I was talking to...good returns on it*': Frontier Economics et al., *The Economic and Social Impacts of Water Trading*, p. 184.

'*Traders do have...their buying power*': Interview with a water trader, November 2020.

'*Picture an entrepreneur...the water baron*': Fullerton [host], 'Sold Down the River', *Four Corners*, 14 July 2003.

'*The VFF never supported...manipulating the water market*': Victorian Farmers Federation, submission in response to ACCC, *ACCC Inquiry into Water Markets in the Murray–Darling Basin: issues paper*, 2019, p. 5.

'*since land and...into agriculture purposes*': Senate Rural and Regional Affairs and Transport Legislation Committee, *Estimates*, 26 May 2017, p. 21.

CHAPTER 20

'*I interviewed Chinese investors...very, very low*': Charlie Huang, quoted in Jan McCallum, 'Why agriculture funding is a hard row to hoe', *BlueNotes*, 22 February 2016.

'*VicSuper is in... through its Kilter fund*': Interview with a water broker, October 2020.

the fund had 4,500 hectares...$430 million in total: James Wagstaff, 'Who owns Australia's farms 2021: the full list of more than 1000 properties', *Weekly Times*, 25 May 2021.

'*is worth about $508 million...Farms of Australia No. 4*': 'Canada is Victoria's biggest water owner', *Weekly Times*, 27 November 2020.

Between 2006 and 2018...to around 48,000 hectares: Marsden Jacob Associates, *Horticulture Below the Choke*, report prepared for the Independent Assessment of Social and Economic Conditions in the Basin, March 2020, p. 29, fig. 17.

And between 1997 and 2018...more than 81,000 hectares: SunRISE Mapping and Research, *2018 Mallee Horticulture Crop Report*, Mallee Catchment Management Authority, 2018, pp. 9, 18, 25.

'*Why would bankers...out of bed for*': Interview with a financial market insider, December 2020.

'*They are planting...how they can trade it*': Interview with a financial market insider, December 2020.

'*The megafarms and... It was a strategy*': Interview with a water market insider, March 2020.

'*There are 7,500 hectares...heard from them*': Frontier Economics et al., *The Economic and Social Impacts of Water Trading*, p. 71.

'*A lot of these...can't do that*': ibid., p. 78.

CHAPTER 21

'*We can always wrap...acting in the market*': Interview with a water trader, November 2020.

CHAPTER 22

'*The crux of the issue...got to pay more*': Lavoipierre and Stockwell [hosts], 'Australia's water barons', *The Signal*, 8 October 2019.

'*hugely higher than...up and up*': Beilharz [host], *Victorian Country Hour*, 26 November 2019.

CHAPTER 23

'*Traders do have...the farmers' walk-away level*': Interview with a water trader, November 2020.

'*With the big investors...to the Middle Ages*': Bart Doohan, in Jérôme Fritel [director], *Lords of Water*, Java Films, 2019.

'*With everyday arbitrage...we take the milk, too*': Interview with a water trader, November 2020.

'*The accounts of banks...tell what is going on*': Interview with a water broker, October 2020.

'*In their 2020 paper...for intertemporal arbitrage*': Seidl, Wheeler and Zuo, 'Treating water markets like stock markets: key water market reform lessons in the Murray–Darling Basin', *Journal of Hydrology*, vol. 581, February 2020.

CHAPTER 24

'*You can't put...a lot less*': Frontier Economics et al., *The Economic and Social Impacts of Water Trading*, p. 70.

'*Water is brought in...a problem for everyone*': ibid., p. 100.

'*One of the things...but it's ridiculous*': ibid., p. 192.

'*This huge place...fads and fashions*': ibid.

'*I can't understand...didn't exist*': ibid., p. 138.

'*There is a lot...drink that every night*': ibid., pp. 201–02.

'*My particular beef...at the regional level*': ibid., pp. 182–83.

'*I'm really cheesed off...to do then?*': ibid., p. 194.

'*Hear that giant...below the Choke*': Interview with a water broker, November 2020.

'*The concern I have...for dairy farmers*': Frontier Economics et al., *The Economic and Social Impacts of Water Trading*, p. 183.

'*For corporate farming...destroys family farms*': ibid., p. 71.

CHAPTER 26

'I had the...more broadly as well': Interview with Tony Windsor, March 2020.

'I've moved a lot of...on communities': Frontier Economics et al., *The Economic and Social Impacts of Water Trading*, p. 183.

'Because of the drought...and be capped': ibid., p. 205.

'Permanent trade within...at severe risk': ibid., p. 137.

'If we don't...in the north': Interview with a water trader, November 2020.

CHAPTER 28

'I'm of the age...to be managed better': Frontier Economics et al., *The Economic and Social Impacts of Water Trading*, p. 193.

'to avoid catastrophic decline...is best managed': AAS, *Investigation of the Causes of Mass Fish Kills in the Menindee Region NSW over the Summer of 2018–2019*, p. 2.

CHAPTER 29

'It is a commodity...of concern to people': Mick Keelty, quoted in Paul Farrell and Alex McDonald, 'Chinese state-owned company buys up water in the Murray–Darling', *ABC News*, 6 May 2020.

'Barriers to trade ... distort market outcomes': ACCC, *ACCC Water Monitoring Report 2009–10*, Australian Government, 2011, p. 97.

'While water market transparency...have evolved': ACCC, *Murray–Darling Basin Water Markets Inquiry: interim report*, p. 324.

'There are currently...to different stakeholders': ibid.

'The 10-point plan...in Australia's history': 'PM's $10b water plan', *Sydney Morning Herald*, 26 January 2007.

'I spent a bit...Water policy is hard': Interview with Malcolm Turnbull, October 2020.

CHAPTER 30

'favours a few and disadvantages many': goFARM Pty Ltd, submission

in response to ACCC, *Murray–Darling Basin Water Markets Inquiry: interim report*, 2020, p. 3.

'*I understand what…buyers to be disqualified*': Interview with Malcolm Turnbull, October 2020.

'*self-determination, self-respect…for quite some time*': Phil Duncan, quoted in Cecilia Harris, '"Nothing about us, without us": call for deeper Indigenous involvement in water catchment management', *Water Source*, 24 August 2020.

CHAPTER 31

Joseph Stiglitz pointed out in a 2014 paper: Joseph E. Stiglitz, 'Tapping the brakes: are less active markets safer and better for the economy?' [conference paper], *2014 Financial Markets Conference*, Atlanta, Georgia, 15 April 2014.

'*Governments allowed competition…participants can trade*': Interview with Ian Shepherd, January 2021.

'*Of course the traders…It's second nature*': ibid.

'*We've forgotten the…enrich smart traders*': Interview with a Basin farmer, October 2020.

'*rigged capitalism*' *and* '*rentier capitalism*': Martin Wolf, 'Martin Wolf: why rigged capitalism is damaging liberal democracy', *Financial Times*, 18 September 2019.

'*To guard against…a common security*': J. M. Powell, *Watering the Garden State: water, land and community in Victoria, 1834–1988*, Allen & Unwin, Sydney, 1989, p. 106.

APPENDIX: TIMELINE

'*For many First Nations…and custodial responsibilities*': Murray Lower Darling Rivers Indigenous Nations, submission to ACCC, *ACCC Inquiry into Water Markets in the Murray–Darling Basin: issues paper*, 2019, pp. 1–2.

'*These forests were…the environment*': Yarkuwa Indigenous Knowledge

Centre Aboriginal Corporation, submission to NSW Natural Resources Commission, *Riverina Bioregion Regional Forest Assessment: river red gum and woodland forests*, 2009, p. 5.

'*Water above all else...the assistance of law*': Plato, *The Laws*, book 8, section 845d, in R. G. Bury [translator], *Plato in Twelve Volumes*, Harvard University Press, Cambridge, and William Heinemann, London, 1967–68.

'*If the water ever...the environment at all*': Maryanne Slattery, quoted in Peter Hannam, 'Barnaby Joyce's department paid "tens of millions" too much for water', *Sydney Morning Herald*, 21 March 2018.

'*pump unless told otherwise*': Anna Henderson, 'The cotton grower, the water minister, the pumping ban and the broken meter', *ABC News*, 14 February 2019.

'*most severe [drought] in 120 years of records*': Kate Doyle, 'Murray–Darling Basin in "most severe" two-to-three year drought conditions in 120 years of records, BOM says', *ABC News*, 19 July 2019.

'*We are now...who has them*': David Littleproud, transcript of media conference in Echuca, 2 September 2019, p. 2.

'*Jackson told* The Fifth Estate...*and other purposes*': Willow Aliento, 'Water and Indigenous people: "I'm tired of being an afterthought",' *Fifth Estate*, 17 October 2019.

'*There is a real concern...droughts and dams*': Richard Kingsford, quoted in Lisa Cox, 'Australia's platypus habitat has shrunk 22% in 30 years, report says', *Guardian*, 23 November 2020.

'*Today was the...accept that compromise*': Lisa Neville, quoted in Kath Sullivan, Clint Jasper and Cassandra Hough, 'Water ministers at odds over best way to progress Murray–Darling Basin Plan', *ABC News*, 28 November 2020.

APPENDIX: WATER WORDS

'*The NWI aims to facilitate efficient water markets*': Productivity Commission, *National Water Reform: issues paper*, Australian Government, 2020, p. 14.